Quality Seed Grower

Quality Seed Grower

Geethu Prabhakar
Aswin Gujrati

2026
Agri Horti Press
New Delhi-110 084

ISBN: 978-93-49690-04-2 (Hardbound)

Published By

Agri Horti Press

New Delhi

E-mail: agrihortipress@gmail.com

Digitally Printed at : Replika Press Pvt. Ltd.

Preface

Quality seed forms the bedrock of successful agriculture, directly influencing crop productivity, resilience, and overall farm profitability. In today's rapidly evolving agricultural landscape, the demand for high-quality, genetically pure, and healthy seeds is more critical than ever.

This book, titled *Quality Seed Grower*, is conceived as a practical and informative guide for farmers, seed producers, agricultural students, and extension professionals who are involved in or aspiring to enter the field of seed production. It aims to bridge the gap between theoretical knowledge and on-ground practices necessary for producing and maintaining superior seed quality.

The early chapters lay out the fundamental principles of seed biology, varietal purity, and genetic maintenance. Readers are introduced to the standards and protocols that govern seed certification, field inspections, and post-harvest handling.

Subsequent sections delve into crop-specific seed production techniques, highlighting isolation distances, rouging practices, pest and disease management, and ideal harvesting methods. The importance of seed testing, storage conditions, and quality assurance mechanisms is also thoroughly discussed.

Special emphasis is placed on modern challenges and opportunities, including climate-resilient seed varieties, farmer-led seed enterprises, and the role of policy and regulation in ensuring seed quality standards. Real-life case studies and illustrative examples have been included to enhance the practical value of the text.

Quality Seed Grower is designed to serve as a field-ready companion and a foundational resource for those committed to the science and art of seed production. It encourages sustainable practices, innovation, and knowledge-sharing, all of which are key to strengthening agricultural systems and ensuring food security.

Geethu Prabhakar

Aswin Gujrati

Contents

Unit 1

Introduction to Quality Seed

1.1 Introduction to seed and seed structure

1.1.1 Seed

The seed in a plant is the part that develops from the ovules after fertilization. They are enclosed in the fruit which develops from the fertilized ovary. The seeds are formed as a result of sexual reproduction and contain the young embryo which can develop into a new plant.

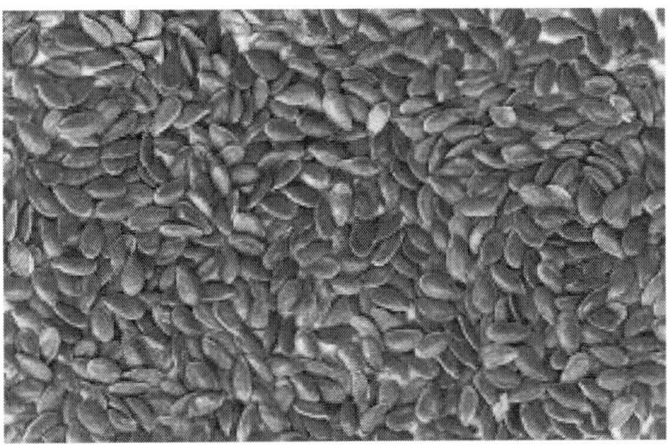

A seed is a basic part of any plant. The ovules after fertilization, develop into seeds. A seed is made up of a seed coat and an embryo. The embryo is made up of a radicle, an embryonal axis and one (wheat, maize) or two cotyledons (gram and pea). A seed is found inside a fruit which converts into a new plant when we plant it. Hence, the seed is the most important part.

There are hundreds of variations in the seed size, shape, colour and surface. The seeds range in size from tiny dust particles, as found in some orchids, to large double-coconuts. The seed surface may be smooth, wrinkled, striate, ribbed, furrowed, reticulate, tuberculate, alveolate, hairy, and pulpy or having patterns like finger prints.

In the seed, life activities are temporarily suspended in order to enable the plant to successfully pass through unfavourable and injurious climatic conditions. On the approach of favourable conditions, the seed resumes active life and grows into full plant. In the form of seeds, a plant can be carried to long distances without special precautions.

1.1.2 Types of Seed

A Seed is primarily of two types:

- ❖ Monocotyledonous Seed
- ❖ Dicotyledonous Seed

Monocotyledonous Seed

A Monocotyledonous seed, as the name suggests, has only one cotyledon. There is only one outer layering of the seed coat. A seed has the following parts:

- ❖ Seed Coat: In the seed of cereals such as maize, the seed coat is membranous and generally fused with the fruit wall, called Hull.
- ❖ Endosperm: The endosperm is bulky and stores food. Generally, monocotyledonous seeds are endospermic but some as in orchids are non-endospermic.
- ❖ Aleuron layer: The outer covering of endosperm separates the embryo by a proteinous layer called aleurone layer.
- ❖ Embryo: The embryo is small and situated in a groove at one end of the endosperm.
- ❖ Scutellum: This is one large and shield-shaped cotyledon.
- ❖ Embryonal axis: Plumule and radicle are the two ends.
- ❖ Coleoptile and coleorhiza: The plumule and radicle are enclosed in sheaths. They are coleoptile and coleorhiza.

Dicotyledonous Seed

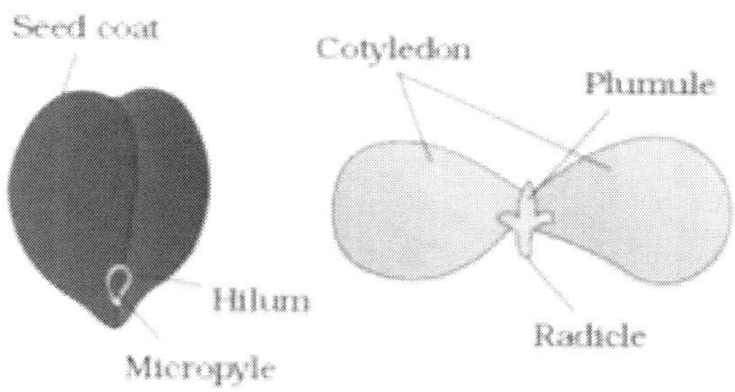

Unlike monocotyledonous seed, a dicotyledonous seed, as the name suggests, has two cotyledons. It has the following parts:

- ❖ Seed coat: This is the outermost covering of a seed. The seed coat has two layers, the outer testa and the inner tegmen.
- ❖ Hilum: The hilum is a scar on the seed coat through which the developing seed was attached to the fruit.
- ❖ Micropyle: It is a small pore present above the hilum.
- ❖ Embryo: It consists of an embryonal axis and two cotyledons.
- ❖ Cotyledons: These are often fleshy and full of reserve food materials.

- ❖ Radicle and plumule: They are present at the two ends of the embryonal axis.
- ❖ Endosperm: In some seeds such as castor, the endosperm formed as a result of double fertilisation, is a food storing tissue. In plants such as bean, gram and pea, the endosperm is not present in the matured seed. They are known as non-endospermous.

1.1.3 Structure of Seed

The various parts of a seed may be easily studied after it has been soaked in water for a day or so varying according to the nature of the seeds. A mature seed contains an embryonic plant (with a radicle and plumule), and is provided with reserve food materials and protective seed coats. A mature pod of pea (Pisum sativum) has a number of seeds arranged in two rows.

The seeds are attached to the fruit wall by a small stalk, the funiculus. At maturity, on one side of the seed coat a narrow, elongated scar representing the point of attachment of seed to its stalk is distinctly seen, this is the hilum. Close to the hilum situated at one end of it there is a minute pore, micropyle. During seed germination, water is absorbed mainly through this pore, and the radicle comes out through it.

Continuous with the hilum there is sort of ridge in the seed coat, the raphe. The seed is covered by two distinct seed coats; the outer whitish one is the testa, while the other inner thin, hyaline and membranous covering is the tegmen. The seed coats give necessary protection to the embryo which lies within.

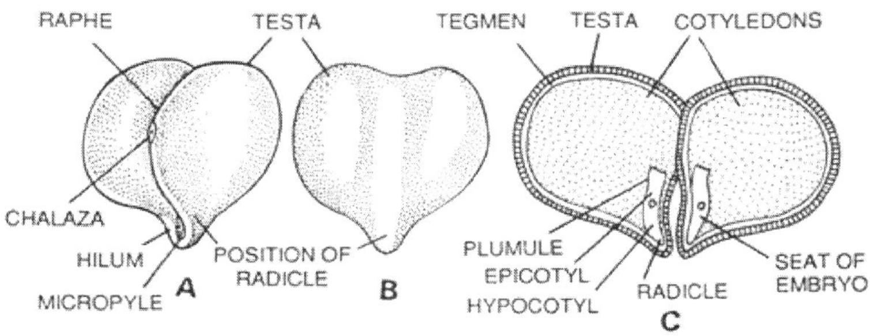

A, complete seed showing various parts; B, seed as seen after removing the testa; C, L.S. of seed.

In this, whitish fleshy body, as seen after removing the seed-coats is the embryo which consists of two fleshy cotyledons and short axis to which the cotyledons remain attached. The position of the axis lying outside the cotyledons, bent inward and directed towards the micropyle is the radicle and the other portion of the axis lying in between the two cotyledons is the plumule.

The plumule is crowned by some minute young leaves. The radicle gives rise to the root, the plumule to the shoot and the cotyledons store up food material. Since the reserve food material is stored in the massive cotyledons and the seed lacks a special nutritive tissue, the endosperm.

The seeds which lack endosperm at maturity are called non-endospermous or exalbuminous. On the other hand in several other plants such as castor bean (Ricinus communis), coconut (Cocos nucifera) and cereals, food is stored in the endosperm. Such seeds where endosperm persists and nourishes the seedling during the initial stages are called endospermous or albuminous.

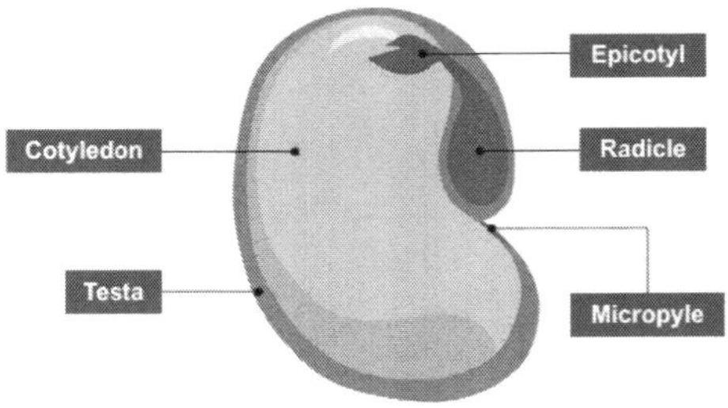

A typical seed will possess the following features:

- Testa – an outer seed coat that protects the embryonic plant
- Micropyle – a small pore in the outer covering of the seed, that allows for the passage of water
- Cotyledon – contains the food stores for the seed and forms the embryonic leaves
- Plumule – the embryonic shoot (also called the epicotyl)
- Radicle – the embryonic root

1.2 Role of quality seed in agriculture

1.2.1 Introduction

Seed is a material which is used for planting/regeneration purpose. It is a fertilized matured ovule which is covered with seed coat known as seed or propagating material which is part of agriculture, sericulture, silviculture and horticultural plants used for sowing or planting.

It is also defined as structurally true seed with fertilized matured ovule having embryonic plant, storage of food and protective seed coat, which has cotyledons and endosperm.

In terms of seed technology, seed may be sexually produced matured ovule having intact embryo, endosperm and or cotyledon with protective covering. It is a propagating material of healthy seedlings, tuber, bulbs, rhizome, roots, cuttings, setts, slips and all types of grafts and vegetatively propagating materials used for production purpose.

It is most vital and crucial input for crop production, as it increase the productivity without adding appreciably to the extent of land under cultivation by planting quality seed.

Good quality seed is essential for production of crop suitable for export. Many farmers produce their own seed, which has been produced by Seed Certification Scheme or use Quality Declared Seed system. Such seed system has been designed to ensure the best possible quality seed production with the resources that are already available in seed production organizations.

The system is designed to provide quality control during seed production which is less demanding on government resources than fully developed seed certification, but adequate to provide good quality seed both within countries and in international trade.

1.2.2 Characteristics of Quality Seed

Quality Seed must be genetically pure having:

- Breeder/nucleus: 100 %
- Foundation seeds: 99.5%
- Certified seeds varieties: 98%
- Certified hybrid seeds:95%
- Certified hybrid cotton: 90%
- Certified hybrid castor: 85%

Quality Seed should have required level of physical purity such as:

- For all crops: 98%
- Carrot: 95%
- Ragi: 97%

Quality Seed should be free from other crop seeds such as designated inseparable crop seeds that are plants seeds cultivated crops found in seed fields where seeds are similar to crop seed that is difficult to separate by mechanical means, causing physical admixtures with crop seeds only when these crop mature at same time when seed crop matures. It covers barley: oats and wheat seeds.

Quality Seed should be free from objectionable weed seeds, as seeds of weed species that are harmful in one or more of following ways;

- size and shape of weed seeds are similar to crop seed and difficult to separate them by mechanical means
- weeds growth habit is determinate type and compete with crop for all resources
- weed plant parts are poisonous or injurious to human and animals
- weed plants serves as alternative hosts for pests and diseases
- Ex. Bersem: chicory
- Paddy: wild paddy (Oryza sativa var. fatua)
- Cucurbits: Wild cucurbits spp.
- Lettuce: wild lettuce
- hendi: wild Abelmoscous spp
- Wheat: Convolvulus arvensis (Hirankuri)

Quality Seed should be free from designated diseases, such as diseases that are specified for seed certification causing contamination of seed lot

- Ex; loose smut of wheat (Ustilago tritici)
- Kernal smut of sorghum (Sphacelotheca sorghi)
- Grain smut of pearl millet (Tolyposporium penicillariae)
- Ergot of bajra (Claviceps microcephala)

- ❖ Early blight of tomato (Alternaria solani)
- ❖ Little leaf of brinjal (Datura virus-2) etc.,

Quality Seed should have high germination and vigour

Quality Seed should have optimum moisture content such as in:

- ❖ Cereals: 10-12 %
- ❖ Pulses: 7-9%
- ❖ Oilseeds:6-7%
- ❖ Vegetables: 5-6%

1.2.3 Importance of Quality Seed:

Seed is a vital input in crop production:

- ❖ It is the cheapest input in crop production and key to agriculture progress.
- ❖ Crop status largely depends on the seed materials used for sowing
- ❖ Response of other inputs in crop production depends on seed material used
- ❖ The seed required for raising crop is quite small and its cost is so less compared to other inputs
- ❖ It emphasis the need for increasing the areas under quality seed production
- ❖ It is estimated that good quality seeds to improved varieties can contribute about 20-25% increase in yield.

1.2.4 Benefits of Using Quality Seeds

- ❖ They are genetically pure.
- ❖ Such seed has high return per unit area as genetic potentiality of crop can be fully exploited.
- ❖ They are less infestation of land with weed seed/other crop seeds.
- ❖ In this, there are less disease and insect problem.
- ❖ It lead to minimize seed/seedling rate i.e., fast and uniform emergence of seedling.
- ❖ They are vigorous, free from pests and disease.
- ❖ They can be adopted themselves for extreme climatic condition and cropping system of the location.
- ❖ Such seeds respond well to applied fertilizers and nutrients.
- ❖ They are uniform in plant population and maturity.
- ❖ Crop raised with quality seed are aesthetically pleasing.
- ❖ Good seed prolongs life of a variety.
- ❖ Yield prediction is very easy.
- ❖ Handling in post-harvest operation will be easy.
- ❖ Preparations of finished products are also better.

- High produce value and their marketability.

1.3 Role of seed growers

- To get the seed information and details about growing seed rules/policies from necessary authorities.
- Should read and follow different growing strategies by discussing with concerned authorities.
- Should be able to know the correct seed information for growing particular seed crop.
- Understand the seed details in terms of purity, usage, growing strategies.
- Growers can get information from state's department of agriculture/Seed Certifying Agencies on where to buy registered seeds.

1.4 Scope and opportunities in seed industries

The Indian seed market has witnessed a major restructuring as a result of the implementation of some progressive policies by the government. Seed Development, 1988 and National Seed Policy, 2002 have helped in strengthening the Indian seed industry in the areas of R&D, product development, supply chain management and quality assurance. Owing to this, India has emerged as the fifth largest seed market across the globe.

Moreover, the active participation of both, public and private sectors has also played a vital role in laying a strong foundation of the industry. This includes launching initiatives to promote the use of hybrid seeds among the farmers who had earlier used outmoded open pollinated varieties. Some other growth-inducing forces, such as growth in income levels, commercialization of agriculture, patent protection systems and intellectual rights over plant varieties, have given a great push to the market.

Substantial increase in yield and quality of crops depends on many factors such as inputs from:

- fertilizers
- irrigation
- plant protection measures
- suitable agronomic practices

The use of high quality seed plays an important role in crop production. The use of poor quality seeds nullifies the utility of all agronomic practices and every other input applied to the crop no matter how lavishly they are applied. Economically, the cost of seed is a very small component of the total cost of production. As seen, importance of seed quality is thus seeds of hope may turn into seeds of frustration when they are not of high quality, so it is important to use seed confirming to prescribe standards in terms of high genetic purity, physical purity, physiological quality and health quality.

Indian farmers were mostly dependent on traditional varieties, so seed requirements were met through farm saved seeds. The use of traditional varieties coupled with farm saved seeds whose quality is not guaranteed, resulted in drastic reduction in production.

Seed is important determinant of agricultural production which depends on performance and efficacy of other inputs. Quality seeds appropriate to different agro-climatic conditions and in sufficient quantity at affordable prices which can raise the productivity. Availability and use of quality seeds is not onetime affair as sustained increase in agriculture production and productivity requires continuous development of new and improved varieties of crops and efficient system of production and supply of seeds to farmers. National Seeds Policy 2002 emphasize that in order to get food production targets of future, main effort is required to upgrade seed replacement rates of many crops which can be done by increasing production of quality seeds which covers thrust areas as:

- Varietal Development
- Seed Production
- Seed Replacement Rate Enhancement
- Primary responsibility for production of breeder seed
- Effective seed production programme
- Popularization of new varieties
- Availability of newly developed varieties to farmers in minimum time gap

1.5 Knowledge about various seed producing organizations

There are three types of Seed Producing Agencies such as:

- Government sector
- Public sector
- Private sector

Public sector organizations include:

National Seeds Corporation (NSC)

NSC was established in 1963 which involves in work of quality seed production. Its responsibilities are defined with the inception of National Seeds Project (NSP) and its main function involves in:

- Production of Breeder Seed: The requirement of breeder seed of various crops is assessed and arrangements are made for its production through crop research institutes of ICAR and State Agricultural Universities.
- Production of Foundation Seed: The production and distribution of the foundation seed of the varieties of national importance is the responsibility of the NSC.
- Production of Certified Seed: The production and marketing of the certified seeds of vegetables, oilseeds, pulses and fodder crops is taken up.
- Marketing: Interstate marketing of the seeds of national varieties lies with the NSC.
- Consultancy Services: The NSC provides consultancy services related to designing, procurement, and installation of seed processing plants, equipments, etc.
- Imparts Training: The NSC organizes short-term training courses to the personnel engaged in seed production work.

- Coordination: NSC provides coordination in the production of certified seed by SSCs.
- Taking up seed certification work in those states where SSCA has not been established.
- Import and export work of seed.

State Seed Corporations (SSCs)

National Seed Project in 1976 establishes State Seed Corporations in 14 states of India. Its functions involve in production and distribution of certified seed in respective state. With the establishment of Tarai Development Corporation in U.P. in 1969 for production of certified seeds, SSCs has been established in other states based on similar pattern to faster the development of seed industry in the country. NSP provides finance for establishment of new SSCs and strengthening of the existing ones. The NSP provides funds for following purposes:

- Strengthening seed production, processing and marketing facilities of NSC.
- Strengthening the facilities of the NSC for producing and processing seeds of vegetables, fodder crops and other specialised crops not taken up by the SSCs.
- Strengthening the facilities for nucleus, breeder and foundation seed production; processing, and storage at selected Central Research Institute of ICAR and at SAUs.

State Seed Certification Agencies (SSCAs): State Seed Certification Agencies are responsible for seed certification in states. SSCAs performs following functions:

- screen applications from seed growers for seed certification and decides on their fitness,
- check and verify appropriateness of source seed used for growing seed crop under certification,
- do requisite field inspections,
- conduct seed tests,
- certify seeds found suitable and issue appropriate tags for certified and foundation seeds,
- guide seed growers on production, processing and distribution of seeds,
- conduct short courses on seed production for seed growers,
- participate in activities conducted to development of seed industry,

Private sector includes various seed companies.

There are many private seed companies which involves in work of quality seed production such as:

- Maharashtra Hybrid Corporation, Jalna, 1964.
- Pioneer Seed Company, Hyderabad.
- Nath Seed Company Pvt. Ltd., Aurangabad.
- Hindustan Lever Ltd., Mumbai.

- Proagro Seed Company, New Delhi.
- Ankur Seed Company, Nagpur.
- Vijay Seed Company, Jalna.
- Mahendra Seed Company, Jalna.

Unit 2

Principles of Seed Production

2.1 Genetic and agronomic principles of seed production

There are two types of principles for seed production which includes:

- ❖ Genetic principles
- ❖ Agronomic principles

2.1.1 Genetic Principle

Genetic purity of seed can be maintained by reducing factors that cause deterioration. Genetic Principle includes:

A. Deterioration of varieties

The genetic purity of variety or trueness to type, deteriorates due to many factors during the production cycles. There are following important factors that are responsible for deterioration of varieties which covers:

Developmental variation:

- ❖ When seed crops are grown under environments with differing soil, fertility, climate photoperiods, or at different elevations for many generation's, then developmental variation sets differential growth responses.
- ❖ It is preferred to grow varieties of crops in areas of natural adaptation to minimize developmental shifts.

Mechanical mixtures:

- ❖ It is an important reason for varietal deterioration which occurs during sowing time when more than one variety is sown with same seed drill by volunteer plants of similar crop in seed field, or by growing many varieties in adjacent fields.
- ❖ In this, two varieties growing next to each other field is mixed at the time of harvesting and threshing operations.
- ❖ In this, threshing equipment is contaminated with seeds of other varieties.
- ❖ Also, gunny bags, seed bins and elevators are contaminated by adding to mechanical mixtures of varieties.

- Roguing the seed fields critically and using utmost care during seed production and processing are required to avoid mechanical contamination.

Mutations:

- It is not a serious deteriorating variety.
- It is difficult to locate/detect minor mutations that occurs naturally.
- In this, mutants are removed by roguing from seed plots to purify seeds.

Natural crossing:

- Natural crossing with undesirable type, diseased plant or off type plant
- It is an important source of varietal deterioration in sexually propagated crops.
- In this, extent of contamination depends on magnitude of natural cross-pollination.
- In this, deterioration sets in due to natural crossing with undesirable types, diseased plants or off types.
- In self-pollinated crops, natural crossing is not serious source of contamination unless variety is male sterile that grows closer with other varieties. The natural crossing, however, can be major source of contamination due to natural crossing.

a) Minor genetic variations:

- It often appears in cross pollinated crops
- It occur in varieties that appears phenotypically uniform and homogenous when released.
- In this, variations gets lost at later production cycles owing to selective elimination by nature.
- In this, yield trials of lines propagated from plants of breeder's seed to maintain purity of self-pollinated crop varieties which can overcome minor variations.
- It requires caring at maintenance of nucleus and breeder's seed of cross-pollinated varieties of crop.

b) Selective influence of diseases:

- In this, new variety easily gets infected with diseases caused by obligate parasites
- The vegetative propagating stocks gets more infected with pathogens
- It requires strict seed production under disease free conditions.

c) ***Techniques of plant breeders***

- Premature release of variety
- Heritable variation due to recombination's and polyploidization occurs in varieties during seed production
- Serious instabilities in varieties owing to cytogenetic irregularities due to improper assessments in release of new varieties.

B. Maintenance of genetic purity

Genetic Purity

Genetic purity is the percentage of contamination by seeds/genetic material of other varieties/species. For any commercial agricultural product propagated by seed begins with the purity of seed planted, so genetic purity of seed planted should be equal/exceed the final product purity standard required, since purity decreases with each subsequent generation of propagation.

It is impossible to assure that no off-type plants or pollen is present in seed production field and that all handling and conveyance equipment and storage facilities are completely free of contamination. The commercial planting seed is 100% pure, so practical seed genetic purity standards established by state seed laws and seed certification agencies ensures the purchaser to receive seed that is in certain purity tolerances.

The tolerances can be achieved based on biology of species, and market-driven standards for final product quality. Earlier generations of seed have strict standards to meet certified seed purity criteria. The main sources of contamination of seed crop are prior crop grown in field, transfer of pollen from nearby field and mixtures during harvesting and handling.

To maintain the genetic purity of variety during seed production, following are the measures which will safeguard the genetic purity, such as:

- ❖ Crop Adoption: To grow crops in the areas that is adaptable to avoid genetic shifts.
- ❖ Approved Seed Class: To use only approved class of seeds in seed multiplication and to adopt generation system.
- ❖ Preceding Crop Requirement: The requirement of crop can be done by inspection and approval of seed plots before planting.
- ❖ Isolation: Isolation of seed crops from many sources of contamination using natural crossing or mechanical mixtures.
- ❖ Roguing: Roguing of off types differing in characteristics from those of seed variety.
- ❖ Field Inspection: It involves qualified and experienced personnel of seed certification agency to inspect seed crops at all required stages of growth and to verify seed lots, purity and quality.
- ❖ GOT: It involves periodic testing of varieties for genetic purity.

There are many steps that suggest for maintaining genetic purity such as:

- ❖ Giving isolation to prevent contamination by natural crossing or mechanical mixtures
- ❖ Rouging of seed fields prior to stage at which they could contaminate the seed crop.
- ❖ Periodic testing of varieties for genetic purity.
- ❖ Avoiding genetic shifts by growing crops in areas in their adaptation only.
- ❖ Certification of seed crops to maintain genetic purity and quality of seed.
- ❖ Adopting the generation system.
- ❖ Grow out tests.

2.1.2 Agronomic principles of seed production

Standardized seed production, besides genetic principles, involves in the application of the following agronomic principles to preserve good seed quality and abundant seed yields.

Selection of suitable Agro-climatic region

- The seed crops have to be grown only in areas well adapted to the photoperiodic and temperature conditions prevailing.
- Region of moderate rainfall and humidity are much more suited to seed production than regions of high rainfall and humidity.
- In general, regions with extreme summer heat and very cold winters should also be avoided for seed production unless particular crops are especially adapted to grow and product under these conditions.

Selection of seed plot

- Seed production plot should have good texture and fertility.
- Should be free from volunteer plants weeds and other crop plants.
- Soil should be free from soil borne diseases and insect pests.
- The previous crops should not be same crop.
- The plot should get adequate isolation distance.

Isolation of seed crops

- The seed crop must be isolated from other nearby fields from the same crop or any contaminating crop as per certification standards.
- Time isolation could also be used in some crops. This is a must to meet the standards for genetic purity of seeds.
- Isolation of seed production of different varieties is also necessary to avoid mechanical contamination.

Preparation of land:

- The land must be prepared well.
- Good land preparation helps in improved & uniform germination resulted in good stand establishment.

Selection of variety:

- Variety should be adopted to agro-climatic condition
- Variety should be high yielder
- Variety should posses other desirable characters like disease resistance, earliness, grain quality etc

Selection of Seed:

- ❖ Seed purity: The seed used for raising a seed crop should be of known purity, appropriate class and invariably obtained from authorized official agency.
- ❖ Seed Treatment: If the seed is not treated already, it should be treated with appropriate fungicides/insecticide prior to sowing.
- ❖ Time of sowing: Should be sown at their normal planting time
- ❖ Seed rate: Lower seed rate than usual to facilitate rogunig operation/seed inspection.
- ❖ Method/depth of sowing: Invariably sown in rows at right depth.

Rouging

- ❖ To rouge plants (off types, pollen shedders, diseased plants, etc.,) at the earliest possible but before flowering
- ❖ The number of rouging varies with the crop, purity of the source seed and the stage of the multiplication of the seed crops.
- ❖ Rouging in most of the field crops may be done at - Vegetative, Pre-flowering, Flowering and Maturity stages

Supplementary pollination

- ❖ In cross-pollinated crops supplementing the natural pollination is very much essential to increase the seed yield.
- ❖ Provision of honeybees, hand pollination, floral mechanism, and rope pulling are some of the techniques used in cross-pollinated crops to increase the seed set and there by seed yield.

Weed Control

- ❖ Compete with seed crop and reduces seed yield and quality
- ❖ Presence of weed seeds at the time of harvest leads to mixing of weed seeds with crops seeds and difficult to separate during seed processing
- ❖ Harbors pests and diseases

Disease and Insect management:

- ❖ Reduces seed yield and quality
- ❖ Infection of seeds with spores

Nutrition:

- ❖ Important role for proper development of plant and seeds
- ❖ N-Good healthy growth P-Fruiting and seed development K- flowering and seed development
- ❖ Proper nutrition to achieve synchronization

Irrigation:

- ❖ Required for translocation of all the nutrients
- ❖ Seed production areas should be dry regions with assured irrigation

Harvesting of seed crops

- ❖ Time and method of harvesting are important.
- ❖ Harvest when the seed is fully matured
- ❖ Early or late harvesting affect seed quality.
- ❖ Every effort should be made to avoid chance of mechanical mixing, mechanical injury to seeds during harvesting, threshing, cleaning drying and packing.

Drying of seeds

- ❖ In order to preserve seed viability and vigour, it necessary to dry seeds to safe moisture content levels as early as possible without heaping wet seed \ pods.

Storage of Raw Seeds

- ❖ After sun drying, the seed should be filled in clean bags.
- ❖ The bags should be dipped in 2 % Melathion, dried and cleaned before they are filled.
- ❖ Mark the bag with name of variety and other details.
- ❖ The bags should be stacked on wood pallets but not directly on the floors.
- ❖ The height of the stack should not be more than 3 to 4 m for cereals, 2.5 to 3 m for other crops.
- ❖ The godowns should be dry, cool and clean and spray with melathion and later fumigate as and when necessary.

2.2 Site selection, season, land requirement, isolation, seed rate, planting ratio, synchronisation, roguing

2.2.1 Site selection

Appropriate site selection is an important factor that determines the success of farm/crop. The site should be easily accessible by road or any form of transport to reach the market for easy disposal. The accessibility of inputs like feed, seed, fertilizer and construction material should also be available nearby the site. While selecting site, grower should make sure that it should be free from pollution, industrial waste, domestic waste and any other harmful activities.

For site selection, ecological, biological and social factors need to be considered.

Ecological factors

Soil

- ❖ Soil quality influences the crop productivity.
- ❖ The properties of soil texture and soil permeability are determined to decide suitability of site.

- Loamy, clay loamy and silt clay soil types are suitable.
- The rocky, sandy, gravel and limestone soil types are to be avoided.
- To produce good seeds, soil must be healthy and fertile.
- The right pH for specific plant should also be obtained.

Temperature

- Temperature has a direct effect on flowering and seed production.
- Tropical plants that flower and produce seeds in hot or tropical areas.
- Temperate plants flower and produce seeds in cold or temperate areas and also produce seeds in cold areas.
- In areas where temperature is not cold, temperate plants can be induced to flower and produce seeds if they are placed in cool conditions before planting.

Water/Rain

- The right amount of water is needed for the growth of plant.
- Hard and continuous rain is not good for seed production since pollen is not transferred; seeds do not develop from flowers; vegetative stage of plant or maturity of seed is prolonged; seeds germinate even if it is not harvested from plant; harvesting becomes more laborious; pests attack or infest the plants and seed yield decreases.
- To prevent seed production during rainy periods, plants can be spaced at wider or longer distances so that all the plants can have enough sunlight.
- In case of less rain or water is not good for plant as it will prevent the normal growth of plant and plant may not produce flowers and seeds.
- Even if flowering occurs, quality of seeds is not good and the seed yield result low.

Wind

- The strength and direction of wind affect the pollination of flowers.

Biological factors

- It include the species to be cultured, seed source and culture type which needs to be considered before site selection of farm.

Social and economic factors

- The ecological and biological factors are prerequisite for good practices in seed site selection and site management.
- It is important to get social and economic background of area and understand the culture and traditions, particularly ideas and beliefs locally associated with agriculture practices.
- The social fabric, market, and its structure, services directly or indirectly linked with seed sector such as transportation, storage, wholesale market aspects etc are to be considered.

- The land identified for farm should be without legal issues and farming should be accepted by local people.
- Other factors include availability of labour, electricity, medical facilities, and transportation.

2.2.2 Land requirement

- The land should be free from crops of previous season of same variety or other varieties of same crop should not have been grown for previous season, if it is same crop, then it should be of same variety that has been certified.
- This selection is highly important for maintenance of genetic purity.
- The land should have adequate irrigation and drainage facilities.
- The land should be free from problem soils as such soil are not suitable for seeds production.
- The pH of the land soil should be between 6-7 as such pH is suitable for good crop.
- The soil should be rich in organic matter and have good water holding capacity.

2.2.3 Isolation

- The crop raised for seed production should be separated from other fields of same crop species by minimum distance, which varies from one crop to other.
- Isolation is essential to prevent pollination from unwanted pollen in case of cross-pollinated and cross-pollinated species to avoid mechanical mixture and chance of cross-pollination in self-pollinated species.
- The larger population size of plants, or larger plot size lead to greater need of isolation.
- Species in same genus/family often have similar minimum isolation distance requirements, but certain varieties in species require larger isolation distances.
- Isolation distance for certified seed production shall be at least 500 feet from varieties of dissimilar adaptation.
- Isolation between classes of same variety may be reduced to 25% of distance otherwise required.
- The crop should have 3 meters of isolation at all sides of seed production plot for maintenance of genetic purity.
- For self pollinated plants, minimum isolation distances range from 50 to 100 feet, as per type of plant.
- Selfers that are easily cross-pollinated by insects require isolation distances from 1/8 mile to 1 mile.
- Outcrossers that are routinely cross-pollinated by insects are typically isolated by 1 mile.
- Crops that have light, wind-borne pollen are often isolated by distances up to 3 miles.
- No isolation is required for production of hand-pollinated seed.
- All cross-pollinating varieties must be 400 feet from any contaminating source.

Isolation distance of different crops

Crop	Minimum isolation distance (m)	
	Foundation seed	Certified seed
Hybrid maize	400	200
Hybrid jowar-other	300	200
jowar -Johnson grass	400	400
Hybrid bajra	1000	200
Wheat	3	3
Rice	3	3
Cotton	50	30
Soybean	3	3
Rapeseed and mustard	400	200
Groundnut	3	3
Sesame	100	50
Peas	20	10
Cowpeas, Sem, Rajma	50	10
Tomato	50	20
Bhendi	400	200
Chillies	400	100
Potato	-	100
Brinjal	200	100
Carrot	1,000	400-800
Cucurbits	800	400
Radish and turnip	1,600	1,000
Cauliflower	1,600	1,000

2.2.4 Seed rate

Introduction

Seeds are the base of agriculture. Technology has modernized much of farming's day-to-day operations, but without steady supply of high-quality seed, yields and crop quality would be greatly decreased. Seed quality also plays an important role in production of agronomic and horticultural crops. Characteristics such as trueness to variety, germination percentage, purity, vigor, and appearance are important to gardeners/farmers in planting crops and to homeowners in establishing their lawns and gardens.

Numbers of plants in a crop depends on seed rate, seed viability, percentage seedling emergence and plant survival. However, plant numbers can often vary widely without appreciably affecting yield in irrigated crops as plants produce tillers, which themselves produce leaves, spikes and grain. There are many sizes and shapes of seed which are from micro-miniature orchid seed, as small as particle of dust, to hat of gargantuan double coconut seed, which is 1 feet long and weighs many pounds. Shapes of seed vary from simple, round tobacco seed to complex, aerodynamic, winged maple seed. Some seeds are hairy like cotton or thistle while others are like grass, have long spikes while some have thorns.

Factors to consider for Seed Rate

The required number of plants per unit area is decided by calculating the seed rate which depends on:

- Plant spacing
- Plant population
- Test weight
- Germination percentage
- Weight of seed
- Plant type; tall or short

Percent Germination

In addition to the number of seed per kilogram, gardeners/growers should also consider percent germination which is proportion of seed that germinates while determining the weight of seed required per acre. As of hereditary and/or seed quality factors, only a part of seed in a given lot will germinate. Percent germination is determined by subjecting samples of specific seed lots to conditions conducive to germination.

The percent germination calculated should be shown on seed label. It is the duty of grower to check the label in order to confirm the percent of seed expected to germinate. If there is any doubt about the percent germination, a germination test should be done before planting the seed.

Weight of Seed

The number of seeds needed for weighing a pound, is known as seed count, which depends on weight of individual seeds. The weight of each seed for a given species is obtained by seed size and specific gravity of seed. Larger seed are heavier, which takes less to weigh a pound. Variation in seed size can have a two-fold effect on recommended seeding rates.

Calculating Seed Rate

To calculate seed rate, he following formula is used:

$$\text{Seed rate (kg/ha)} = \frac{\text{Plant population (per ha)} \times \text{No. of seeds/hill} \times \text{Test weight (g)} \times 100}{1000 \times 1000 \times \text{Germination percentage (\%)}}$$

2.2.5 Planting ratio

- It is the ratio in which male and female parental lines are planted in hybrid seed production or maintenance of cytoplasmic male sterile line.

2.2.6 Roguing

- It is the removal of plants which are off type, phenotypically different from the plants of the variety under certification.
- It is an important aspect of seed production and is necessary to prevent outcrossing and mechanical mixture.
- In this, off-type plants are regularly removed from the field either by uprooting or by cutting at the ground level.
- The off-type plants may differ in plant height leaf characters, flowering time, maturity etc.

2.2.7 Season

- Seed crop should be grown in appropriate season as seasonal difference will affect the quality of seed
- For quality seed production, dry and cool weather is more favourable than hot and humid weather during panicle emergence and seed development stage.
- Rains during seed maturation causes seed discolouration and ergot disease.
- For seed production, sowing to be done during June – July and October – November seasons where October-November is more conductive for quality seed production.

2.3 Important steps for hybrid seed production

2.3.1 Introduction

The seed production involving first generation progeny crossing between two genetically diversified parents is called as hybrid seed production. It involves multiplication of cytoplasmic genetic male sterile line (A line) , maintainer line (B line) and restore line (R line) and production of F1 hybrid seed

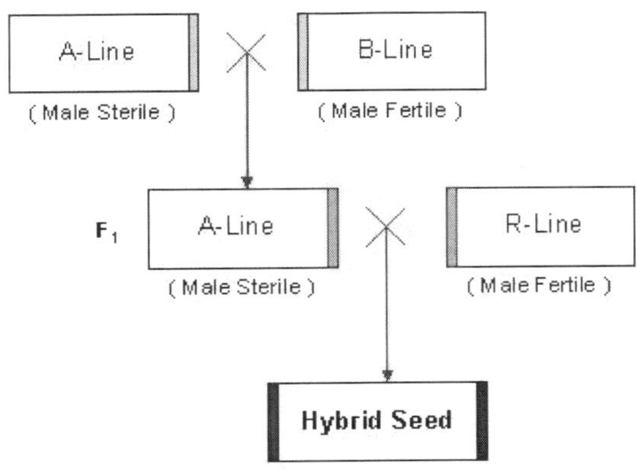

Hybrid is produced by crossing between two genetically dissimilar parents where pollen from male parent will pollinate, fertilize and set seeds in female to produce F1 hybrid seeds. For production of hybrid crossing between two parents is important, which results in heterosis.

2.3.2 Requisites for hybrid seed production

Breeders responsibilities

(a) Develop inbred lines

(b) Indentification of specific parental lines

(c) Develop system for pollen control

Major problems for breeders & producers

(a) Maintenance of parental lines

(b) Separation of male and female reproductive organs

(c) Pollen exchange

Genetically it makes no difference which parental line is used as the male or female; but seed producers must consider the following characteristics of parental lines.

Female Parent	Male Parent
High seed yield	Good pollen production
Good seed characteristics	Long shedding period
Pest resistance	Plant height
Male sterility	Fertility restoration
Lodging resistant	

Procedures for hybrid seed production

1. Development and identification for parental lines
2. Multiplication of parental lines
3. Production of single crosses

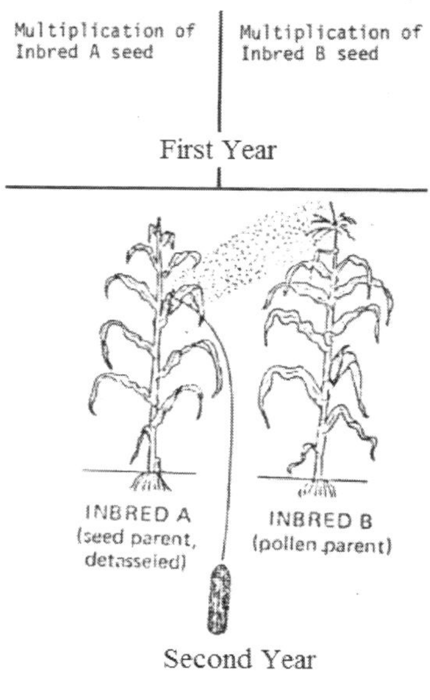

(a) Planting ratios

(b) Planting dates

4. Production of double cross hybrids

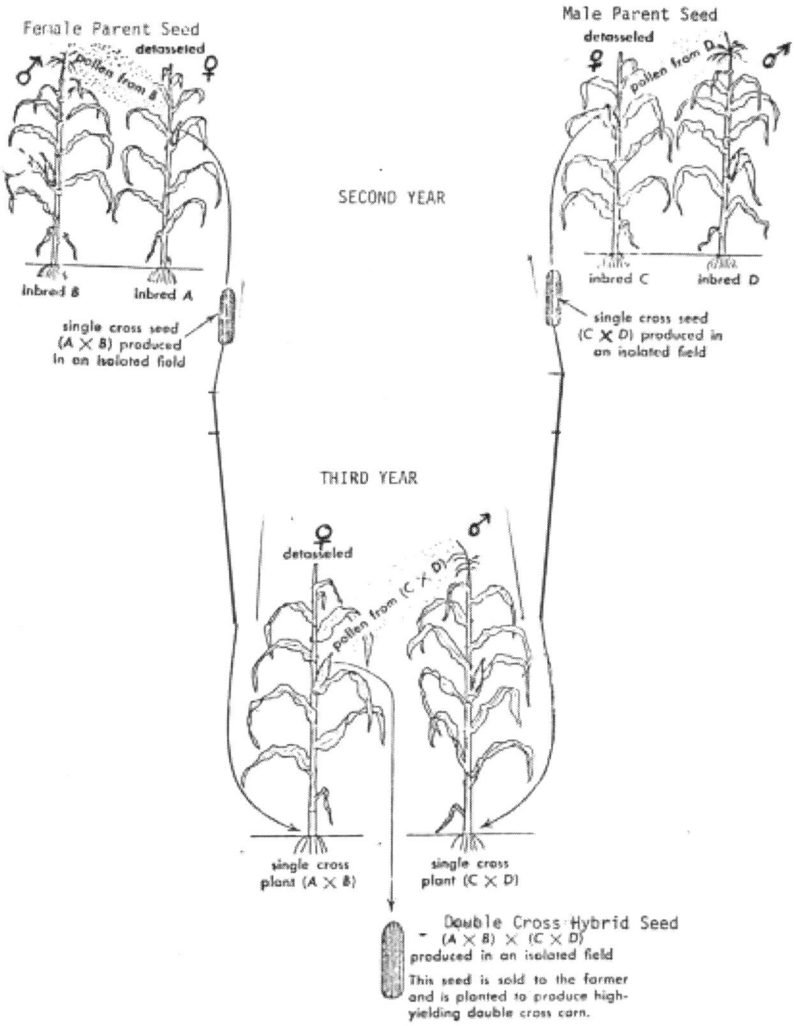

Using male sterile and restoration factors

- Male sterility does not occur frequently in natural populations as isolated male sterile plant cannot reproduce itself i.e. male sterility is dead-end in nature.
- Male sterility is useful in development of superior genotypes and seed production of hybrid varieties.
- Control of male sterility

2.3.3 Agronomic Approaches

There are many strategies that are used for hybrid seed production such as:

- Site selection: Medium to heavy deep well drained soil is good for seed production.
- Isolation distance: Maintenance of genetic purity for certified seed production of male sterility as per hybrids depends on proper creation of isolation distance. The isolation distance between parents should be minimum of 5 m.
- Fertilizer requirement: The fertilizer dose of 100: 50: 50 or 150: 75: 75 kg/ha NPK as per soil fertility, out of which 50: 50: 50 or 75: 75: 75 NPK as basal dose and remaining 50 or 75 (N) is applied in two splits as at starting of square formation and at boll development.
- Raising of parent lines: If male sterile lines are used as parent's lines, then certain steps/ rules are adopted for raising of parental lines, which includes inbred line approval by certified agency, progeny of 2 approved inbred lines of which one is male sterile, inbred line to be true breeding strain resulting from self-pollination with selection.
- Sowing: Sowing date of parental lines are so adjusted that there is synchronization of flowering in females and male parent and there will regular supply of pollen till crossing season gets over. Staggered planting of male done as per date of flowering in male and female.
- Plant protection measures: Attack of sucking pest is controlled by confidor @ 100 ml/ ha. Boll worm attack on flowers and squares are manage by spraying endosulphan @ 2L/ha.
- Irrigation scheduling: Use of alternate furrow method and irrigation to be done whenever required.
- Physiological disorder: Leaf reddening, potassium deficiency and boll shedding are main disorders related to physiology of hybrid.
- Picking: Picking of bolls that are opened and collect it in proper bags.
- Storage: Clean and store in gunny bags with proper tagging. Under developed and damaged bolls should be removed and the remaining should be sent for ginning purpose and finally hand over to seed certification.

2.3.4 Steps in Hybrid Seed Production

There are certain emasculation methods that needs highly skilled labour, which can be adopted as per flower types. The flower buds that are likely to open the next-day are selected for emasculation, for which best time is afternoon. This can be done by either of the methods:

Doak/thumb nail method

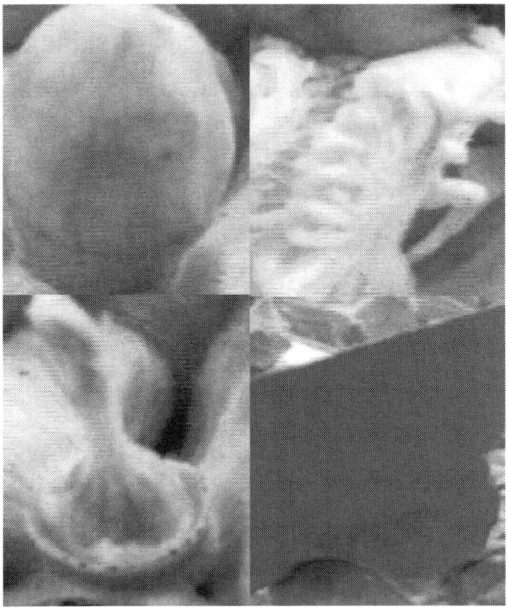

- ❖ It is a successful method which is used in hybrid seed production.
- ❖ It involves removal of corolla with anther sheath by shallow cut at base with thumb nail.
- ❖ In this, emasculated flower buds are covered with paper bag to prevent contamination from foreign pollen.
- ❖ In this, if no open flowers are left over in field of female parent, bagging may be avoided but marking of emasculated flower helps in identification during pollination.
- ❖ This method is not suitable for developing diploid or desi hybrids as flower buds of these are small and style is short and brittle which makes the method unsuitable in large scale seed production of hybrid seeds.

Pinching off top of corolla:

- ❖ The method is used for emasculation of diploid flowers where flower top bud is pinched loss using thumb nail and first finger nails by hand so that stigma tip get slightly exposed and bud is covered with mud, as bud matures, the stigma tip extend sufficiently for enabling pollination.
- ❖ In this, as the buds mature, stigma tip extend sufficiently to allow pollination.
- ❖ In genotypes where stigma tip exsertion is low, then instead of cutting top portion of corolla, entire corolla is removed and lint dipped in mud is applied on unopened anther sacs.
- ❖ It is an easy method of emasculation for diploid flowers as chances are there which leads to selfed seed production.

Straw tube method:

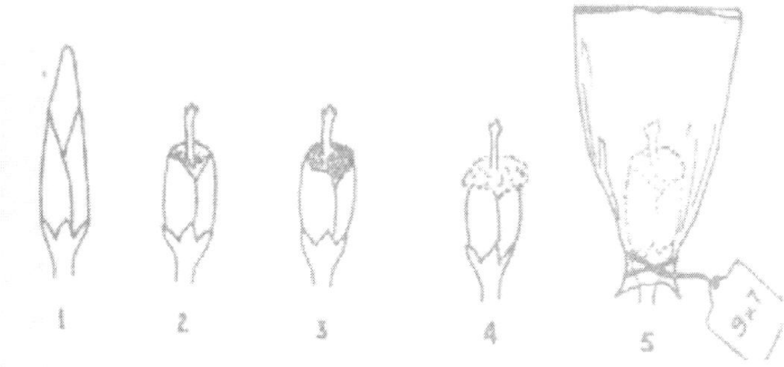

- ❖ In this, top of corolla of bud is pinched off and piece of straw is used by inserting in style to separate all anthers from anther column and leaving the tube in same position.
- ❖ In this pollen is applied in the next morning.
- ❖ During emasculation, more time is required for inserting straw tube exactly on style so that anthers would get dislodged.

Petal remove and brushing off anthers:

- ❖ In this, petals are removed by thumb nail.
- ❖ The anthers brushed off by lightly touching and moving thumb on stamina column.
- ❖ The method is useful for herbaceous type of buds as anthers are granule shape and easily removable.

2.4 Testing, release, notification and maintenance of varieties/hybrids

2.4.1 Testing

In growth of agriculture, improved varieties of crop plays an important role in development, so main objective of crop science research involves in development of improved varieties in different crops. The new improved varieties of crops are developed by Crop Research Institutes or State Agricultural Universities or by Private Seed Companies.

The crop varieties are tested for minimum period of three years, before consideration of release for cultivation. The important characters for which new varieties are evaluated/ tested include:

- ❖ Yield
- ❖ Resistance to disease and pests
- ❖ Quality of product
- ❖ Resistance to adverse environmental factors
- ❖ Adaptation to different agro climatic conditions

2.4.2 Variety release

Norms for release of New Variety:

There are certain norms that should be considered for release of new variety. Besides distinctness, uniformity and stability, the new variety should be superior to previous released variety in one or more of following characteristics:

- Yield potential of grain, oil, fibre, fodder, vegetable or other economic product.
- Resistance to biotic and abiotic stresses.
- Quality of oil, fibre, protein, fodder, vegetable, grain, etc.
- Maturity duration.
- Suitability for general cultivation over a wide range of environmental conditions.
- Suitability for machine harvesting, etc.

Types of varieties:

Pure-Line Variety:

- It is developed in self-pollinated crops by pure-line selection.
- The progeny of single homozygous plant obtained by selfing is called as pure-line.
- The variety is released from single best progeny.
- The base population, which is used for development of pure-line variety is always a heterogeneous population.
- A pure-line variety is a homozygous and homogeneous population.

Mass Selected Variety:

- It is developed in self and cross pollinated species.
- In this, mass selection is more common in cross pollinated crops than in self-pollinated crops.
- In self- pollinated crops, a mass selected variety consists of several pure-lines which is a mixture of several pure-lines.
- In cross pollinated species, a mass selected variety is a mixture of several homo and heterozygotes, that can be heterozygous and heterogeneous population.
- A mass selected variety has broad genetic base and wide adaptation.

Clonal Variety:

- It is developed in asexually propagated crops like potato, sugarcane, sweet potato, etc.
- The progeny of single best clone is released as a cultivar.
- Such varieties are heterozygous but homogeneous, as all plants of clone have exactly the same genotype.
- Clonal cultivars have wide adaptation than pure-line cultivars.
- A heterogeneous base population is used for development of clonal variety.

Synthetic Variety:

- ❖ A synthetic variety is developed in cross pollinated crops such as maize.
- ❖ For development of synthetic variety, good general combining inbreds are identified.
- ❖ In this, inbreds with good general combining ability are intermated in all possible combinations and hybrid seed is obtained.
- ❖ The hybrid seed of all combinations is mixed together in equal quantity to constitute a synthetic variety.
- ❖ In this, synthetic variety has several heterozygotes which is maintained by open pollination to allow some degree of selfing.
- ❖ In later generations, synthetic variety has several homo and heterozygotes, so synthetic variety has heterogeneous population and wide adaptation.
- ❖ The yield level of synthetic variety is always higher than open pollinated variety but lower than the yield of single and double cross hybrids.

Composite Variety:

- ❖ It is developed in cross pollinated species.
- ❖ It is developed by mixing the seed of various open pollinated genotypes which are similar in height, maturity, seed colour, seed size, etc.
- ❖ The composite variety has heterozygous and heterogeneous population.
- ❖ It consists of many homozygotes and heterozygotes.
- ❖ The varieties are highly adaptable to environmental fluctuations.

Multiline Variety:

- ❖ It is developed in self-pollinated crops.
- ❖ A multiline variety is a mixture of several isogenic lines, related lines or unrelated lines.
- ❖ In this, each component line has different gene for disease resistance.
- ❖ Such varieties are homozygous but heterogeneous and have very good buffering capacity due to broad genetic base.

Mutant Variety:

- ❖ A mutant variety is developed through induced mutagenesis.
- ❖ Mutant varieties are developed in self-pollinated and asexually propagated species and rarely in cross pollinated species.
- ❖ A mutant variety differs from parent variety in one or two characters.
- ❖ Such varieties are developed to induce disease resistance, earliness, dwarfness and achieve improvement in quality.

Transgenic Variety:

- ❖ A variety which is developed by the technique of genetic engineering is called transgenic variety.

- Such varieties are developed to solve specific problems that are difficult to be solved by conventional methods of plant breeding.
- Transgenic varieties with herbicide resistance, insect resistance and improved quality have "been developed in several crop plants.
- Such varieties can be developed in all the three groups of crops viz. self-pollinated, cross pollinated and asexually propagated.

Pedigree, Bulk and Backcross Bred Varieties:

- Pedigree and bulk bred varieties are developed in self-pollinated species.
- These varieties are developed from progeny of a single best homozygote.
- Varieties developed by pedigree and bulk methods are homozygous and homogeneous and have narrow genetic base.
- Backcross method is used both in self and cross pollinated species for improvement of oligogenic character such as disease resistance.
- Backcross bred variety is similar to parent variety except for the gene which has been transferred into parent variety.

Hybrid Variety:

- The F1 population that is used for commercial cultivation is hybrid variety.
- Hybrid varieties are developed in cross and self-pollinated species.
- Hybrid varieties are more common in cross pollinated species than in self-pollinated species.
- Hybrids are of different types, viz., intervarietal hybrids and interspecific hybrids.
- Hybrids are heterozygous but homogeneous.
- For hybrid varieties, fresh seed has to be produced every year.

Features

- Variety release is main stage in formal seed chain which links plant breeding to farmers.
- It covers all procedures and actions that takes place from time when breeder decides to enter a promising new line/variety for official testing till the time when new variety is made available for multiplication as first step on road to general cultivation by farmers.
- Release relates to physical activities linked with multiplication program or promotional campaign.
- In public sector varieties, registration is followed directly by release, provided sufficient early generation material to initiate multiplication.
- In this, public breeders are judged on their output of varieties and are reluctant to use limited resources for maintenance and seed multiplication.
- The new varieties should be marketed once registered.

- ❖ For registering, tests for distinctiveness, uniformity, stability and value for cultivation and use needs to be conducted first before registration.
- ❖ The registration establishes legal ownership of the new variety.
- ❖ The seed laws on variety testing and release variable and inconsistent.
- ❖ The variability and inconsistency of seed laws makes it costly for seed companies to release and market new varieties.
- ❖ A new variety must be tested each time it is to be marketed in respective countries, even if it is developed for sale across wide range of agro-ecologies.
- ❖ In many cases the variety was not released based on its merit and uniqueness.

Process of Variety Release

- ❖ To release of new crop varieties, it covers:
- ❖ Development of new strains,
- ❖ Evaluation of performance,
- ❖ Identification of superior strains,
- ❖ Release and notification.

2.4.3 Notification

- ❖ Notification is a statutory function which is performed under Seeds Act so that the provisions of Seeds Act could be applied to regulate the quality of seeds during the sale.
- ❖ The notification is made by Central Government on recommendation of Central Seed Committee.
- ❖ The proposals for notification of State-important variety are forwarded in prescribed format by State government after its release in State to Central Seed Committee for consideration .
- ❖ In this, cultivars are notified under Seeds Act so that quality of seeds can be regulated.

Purpose

The main purpose of notification is to:

- ❖ bring the seeds of particular crop/variety under the purview of Seed Law Enforcement
- ❖ empower the Seed Inspectors to verify the quality of its seeds by sampling and analysis

Advantages

- ❖ Notification is compulsory for production of certified seeds.
- ❖ Unless the variety is notified, the seed cannot be certified.
- ❖ For regulation of quality of seeds, notification is a precondition.
- ❖ As specified under Seeds Act, seeds of notified varieties can be sold after proper labeling and packing indicating the minimum specified standards.
- ❖ Once the variety is notified, it becomes government asset.

- The morphological character of notified varieties are documented by Central Seed Committee so as to curtail the bio-piracy.
- Farmers get seeds of assured quality of notified varieties.
- Help to trace out its genesis.
- To undertake Seed planning/programmes

2.4.4 Maintenance of variety

Features

- Maintenance of variety in cross pollinated crops is much more complicated than self pollinated crops.
- Maintenance of nucleus seed of inbred lines is done once the hybrid has thoroughly tested and if suitable, seed of parental lines increases.
- Hand pollination involves method of maintaining nucleus seed of inbred lines by self pollination, sib pollination of combination of both.
- Maintenance by sibbing is preferred as it will not reduce vigour and maintain parental material by alternate selfing and sibbing from one generation to next.

Maintenance of Genetic Purity

The following methods have been suggested for maintenance of genetic purity:

1. Use of approved seed in seed multiplication
2. Inspection of seed fields prior to planting
3. Field inspection of seed crops
4. Sampling and sealing of cleaned lots
5. Pre/post control

Genetic purity maintenance

Various steps suggested for maintaining genetic purity are:

1. Providing isolation to prevent cross fertilization or mechanical mixtures
2. Rouging of seed fields prior to planting
3. Grow in adapted areas only to avoid genetic shifts in the variety
4. Certification of seed crops to maintain genetic purity and quality
5. Adopting generation system

Procedures for variety maintenance

Maintenance procedures are the extension of normal breeding process but selection is mild and aims not to improve the variety by to keep the identity unchanged. The commonly used produces are

- Mass selection
- Ear-to-row/Plant-to-row

Maintenance procedure for self-pollinating crops

The maintenance procedure for selfpollinating crops involves pure line or ear-to-row selection method which involves:

- ❖ At least 100 true-to-types ears or plants are selected and harvested separately
- ❖ Seeds from each plant are planted together in separate rows or small plots
- ❖ Eliminate rows/plots which do not conform to variety description
- ❖ Only row/plots that are uniform and definitely true to variety are harvested for seed
- ❖ Harvested seed from different uniform looking rows that can be bulked to constitute breeder seed

2.5 Indian Minimum Seed Certification Standards

2.5.1 Introduction

- ❖ General Seed Certification Standards are applicable to all crops which are eligible for certification, and with field and seed standards for individual crops, constitutes Minimum Seed Certification Standards. The idea of seed certification is to maintain/ aware to public, by certification of high quality seeds and propagating materials of required kind and varieties of grown and distributed seeds to ensure genetic identity and genetic purity. It is also designed to achieve prescribed standards.
- ❖ The minimum seed certification standards, thus are the standards required for the certification of seeds by the certification agencies.
- ❖ Minimum seed certification standards are grouped as General Seed Certification Standards and Specific Crop Standards.

2.5.2 Phases of Seed Certification

Certification of seeds can be done in following phases:

- ❖ receipt and scrutiny of application
- ❖ verification of seed source, class and seed requirements for raising seed crop;
- ❖ field inspections to verify conformity to prescribed field standards;
- ❖ supervision at post-harvest stages covering processing and packing;
- ❖ seed sampling and analysis, covering genetic purity test and/or seed health test, to verify conformity to required standards;
- ❖ granting certification, tags, tagging and sealing

2.5.3 Process involved

- ❖ Administrative checking of origin of propagating material
- ❖ Field Inspection
- ❖ Sample inspection
- ❖ Control plot testing
- ❖ Grow-out test

2.5.4 Minimum field standards

Crop	Class of seed	Land requirement	Isolation (meters)	No. of field inspections	Off types	Inseparable other crop plants	Objectionable weed plants	Affected Plant /head
Paddy	Foundation Seed	--	3	2	.05	--	.01	--
	Certified Seed	--	"	2	.20	--	.02	--
Wheat	Foundation Seed	--	3,	2	.05	.01	--	.10
			150*					
	Certified Seed	--		2	.20	.05	--	.50
Sorghum (open pollinated varieties)	Foundation Seed	--	200, 400*	3	.05** .10***	--	--	.05
Pulses	Foundation Seed	--	10	2	.10	--	--	--
Oilseeds Groundnut	Foundation Seed	Two season	3	2	.10	--	--	--
Soybean	Foundation Seed	--	3	2	.10	--	--	--
	Certified Seed	--	3	2	.50	--	--	--
Fiber Crops Cotton	Certified Seed	--	30,5*	2	.20	--	--	--
Vegetable Crops Potato Seed	Certified Seed	*	50,	4	.10	--	--	3.0

Crop	Class of seed	Land requirement	Isolation (meters)	No. of field inspections	Off types	Inseparable other crop plants	Objectionable weed plants	Affected Plant /head
Tomato	Foundation Seed	--	50	3	.10	--	--	.10
	Certified Seed	--	25	3	.20	--	--	.50
Cauliflower	Certified Seed	--	1000*	3	.20	--	--	.50
Onion	Foundation Seed	--	5	2*	.10	--	--	--

2.6 Seed certification system, seed act and rules to maintain seed quality

2.6.1 Seed certification system

Introduction

Seed certification plays an important role in production of pedigreed seeds of many agricultural crops. Since inception, seed certification has grown steadily to its present impressive level. Throughout its growth, seed certification has been continually challenged to improve and expand its services, to keep abreast of changes in plant breeding, seed marketing, and farmer interest.

Seed certification is legally sanctioned system for quality control of seed multiplication and production. The idea of seed certification is to maintain and make available to public, through certification, high quality seeds and propagating materials of notified kind and varieties so grown and distributed as to ensure genetic identity and genetic purity.

It is a quality assurance system where seeds are intended for marketing subjected to official control and inspection. The system certifies that sack, packet or box of seed contains as per the label and that seed was produced, inspected and graded, as per the needs of Certification Scheme.

The objective of seed certification is to supply high quality seed to farmers and other growers, which is true to identity, high in purity and germination capacity and free from certain pests and diseases. Seed quality is most important in crop production, as high quality seed is essential for good crop yields and good returns, and minimises the likelihood of crop failure.

It is an official system which ensures that seed is produced, multiplied and marketed as per predetermined standards and systems while maintaining the genetic integrity of the product. It is part of a wider system which covers plant breeding and plant breeder's rights, plant genetic resources, biodiversity and international trade.

Concepts

Seed Certification express following fundamental concepts:

- The integrity of certified seed growers must be recongnised.
- Field inspection must be made by qualified inspectors.
- Verification trials to establish identification and usefulness of varieties and strains certified must be conducted.
- It must recognised that there is necessity for keeping proper records to establish and maintain satisfactory pedigree of seed stocks.
- There should be standards established for purity and germination.
- The principle of sealing seeds to protect both grower and purchaser must be approved.
- Species of farm weeds which would be included within the meaning of noxious weeds, must be defined.

Phases

- Verification of seed source
- Receipt and scrutiny of application/Farm-1
- Field inspection
- Supervision of post harvesting and Seed Processing upgrading operations and sampling
- Seed Quality analysis and genetic purity evaluation
- Seed Treatment
- Bagging and Tagging
- Grant of Certificate
- Issue of release order (Farm-II)

Criteria

- The seed varieties which are specified under Section 5 of Seeds Act, are eligible for seed certification.
- Should be notified variety under Section 5 of Indian Seed Act.
- Should be in production chain and its pedigree must be identifiable.
- Should possess minimum field standards including selection of site, isolation requirements, spacing, planting ratio, border rows etc.
- Presence of off-types in seed crop, pollen-shedders in Sorghum, Bajra, Sunflower etc., Shedding tassels in maize crosses, the disease affected plants, objectionable weed plants should be in allowed levels while applying for seed certification.
- Maximum seed certification standards are required for each crop.

Steps in Seed Certification System

The following are the steps involved in seed certification:

- Evaluation of origin: The source seed verification is initial step in Seed Certification. Unless seed is from an approved source and designated class, the certifying agency will not allow seed field for certification, thereby ensuring use of high quality seed for sowing crops.
- Field Inspection: Field inspection involves evaluation of crop in field for varied purity, isolation of seed crop to prevent out-cross, disease dissemination and to ensure crop condition about presence of objectionable weed plants and designated diseases etc.
- Sample inspection: With lab tests, value of seeds for planting will be assessed and after that the certifying agency holds samples from seeds provided under certification programme and further allows to undergo germination and other purity tests for conforming varietal purity.
- Bulk Inspection: In certification, provision has been made for bulk inspection. The evaluation to check homogeneity of bulk seed produced as compared with standard sample is carried out.

- Control plot testing: In this, samples were drawn from source and final seed produced are allowed to grow along with standard samples of variety. On comparison, results can be obtained whether varietal purity and health of seed produced are equal to results on field inspection.
- Grow-out test: It involves finding seeds for their genuineness to species or varieties or any identification of infected seeds. Here samples from the fields are grown along with regular checks where growing plants are observed for varietal purity. Such test helps in removing of sub-standard seed lots.

2.6.2 Seed act

Introduction

In 1966, parliament had passed the Seeds Act to:

- provide legal framework around seed certification
- to make good quality seeds available to cultivators
- frame and notified Seed rules
- start systematic Seed Certification
- establish Central Seed Committee in respective states
- provide establishment of Seed Certification Agencies in states; Seed certification Boards and State Seed Testing Laboratories.

Clauses

The Seeds Act 1966 has total 25 clauses which are summarized below:

- It covers seeds of food crops, oil crops, cotton seeds, seeds of cattle fodder and all types of vegetative propagating material.
- It provides for establishment of Central Seed Committee to advise Central and State Governments on matters arising out of the administration of this act and carry out other functions assigned to it by the Act.
- It provides establishing a Central Seed Laboratory as well as State Seed Laboratory to carry out seed analysis of notified variety.
- It empowers Central Seed Committee to notify any variety found suitable as per the Act after notification in Official Gazette.
- It empowers committee to fix the minimum limits of germination and purity of seed for a variety to be notified as well as for marking or labelling a seed lot to be sold commercially.
- Regulation of sale of seeds of notified varieties by compulsory truthful labeling revealing the true identity of the variety, germination as well as purity.
- Constituting certification agency for undertaking process of certification.
- Power of certification agency to recommend notification of suitable variety and grant of notification certificate provided the seed meets minimum limits of germination and purity.

- Empowerment to agency for revocation of certificate if agency is convinced that holder has obtained certificate by misrepresentation or not complied with the conditions.
- Provision for an appeal by holder on payment basis to express before an appellate in India authority, his limitations for not complying with the conditions.
- Appointment of seed analyst to undertake seed testing.
- Appointment of seed inspector who is a public servant in meaning or section 21 of Indian Penal Code.
- Empowerment of seed inspector to draw samples from any seller or purchaser and verify quality by sending samples to seed analyst in seed testing labs.
- Follow procedure for seed sample collection and other rules.
- The clause entrust inspector with power to break open any seed container or door of any premises where seed may be kept for sale, under circumstances when owner refuses to cooperate.
- Responsibility of Seed analyst to report results in specific format after analysis of seed samples to Seed Inspector and seller/ purchaser.
- Restriction on import and export of seeds of notified varieties.
- Recognition of seed certification agencies of foreign countries for purpose of this act.
- Penalty or punishment or both for those who do not comply with provisions of act and to prevent seed inspectors from executing his power
- Forfeiture of property/seeds belonging to any person convicted under this act due to contravention of procedures under this act.
- Punishment for offences committed by companies or anybody corporate.
- Protection of Government action taken in good faith that is no prosecution or legal proceeding will lie against Government or any Government Officer for anything that is done in good faith.
- Power for Government to give directions for smooth conduct of act 24.
- Power of Government to make rules to carry out various functions of Central Seed Committee, Central Seed Laboratory, Certification Agency and Seed Inspectors.

Unit 3

Seed Production Management in Fields

3.1 Field history and field selection & preparation

3.1.1 Field history for field selection

Field history for seed production

It is important for seed grower to have good history record for the field where he/she will grow their crops. Knowing about previous information of field will be useful for agronomy field scout. Such information is also helpful when a seed grower or farmer has complete record about the field which covers information related to:

- legal land location
- field name
- history of previous crops
- soil fertility problems
- nutrient deficiencies
- previous problems with insects
- past weed types and populations
- herbicide-resistant weeds
- previous disease problems
- previous herbicides and herbicide groups used
- crop yields

For current crop year, seed grower should list the specific crop variety, seeding date, seeding rate, fertilizer used and tillage operations.

Field Inspection

Previous history can be very important when assessing crop problems. To know about the fields for respective seed production, field inspection is required in order to verify conformity of seed crops to prescribed certification standards which should have information related to:

- ❖ Seed origin (source seed) and identity of variety.
- ❖ Cropping history of seed field.
- ❖ Crop and cultivation conditions.
- ❖ Isolation distance.
- ❖ Freedom from impurities, crop plants and weeds plants.
- ❖ Freedom from other cultivars, off-types and seed borne diseases.

Information related to contaminants

To get the details about the field, seed grower should know about various contaminants which needs to be observed during field inspection, which are classified as:

1. Off types

- ❖ Plants of same crop species as seed crop, differing in expression of morphological characters like plant type, branching type, pigmentation etc. are classified as off types.
- ❖ To designate plant as off type, it is not necessary to identify it definitely as of another variety.
- ❖ While taking field counts, off type is counted as off type irrespective of its stage of growth.
- ❖ When the stage of growth of seed crop and off type differ so widely that off type cannot contaminate the seed crop at any stage, genetically or physically, then such off types should be made so that observation is complete.
- ❖ For field counts, off types need not to be counted.
- ❖ The seed grower should be informed of presence of such off types so that he/she may removed.
- ❖ Off types not causing contamination at time of recording field counts, but likely to cause contamination later must be counted.

2. Inseparable other crop plants

- ❖ Inseparable other crop plants are plants of cultivated crops found in seed field whose seeds are similar to crop seed, which is difficult to separate economically by mechanical means.
- ❖ An inseparable other crop plant is counted if its stage of growth is such that it would bear seed when the seed crop matures, and causes mechanical admixture during harvesting and threshing.
- ❖ If stage of growth is different from that of seed crop which cannot bear seed by time, seed crop need not be counted.

3. Objectionable weed plants

- ❖ Plants of weed species whose seed size and shape are similar to that of crop seeds and which are difficult to remove from seed economically by mechanical means.

- ❖ Weed species whose eradication is difficult if once introduced, or those who serve as alternate hosts for crop pests or diseases, are objectionable weed plants.

4. Diseased plants

- ❖ The plants affected by designated diseases should invariably be counted.
- ❖ The grower should be advised to rogue them out from seed fields.

Selection of field

- ❖ Field selection shall be done in coordination with seed production cooperative.
- ❖ For seed production, clean, plain, fertile land with good drainage facility should be selected, where seed grower needs to consider following things:
    - Seed production field should be fertile land with easy facility of irrigation and drainage. There are certain seed crops that generally does not need irrigation, however irrigation is necessary at tillering and flowering stage.
    - Easily accessible for supervision or field inspection.
    - The field selected for seed production should not be diseased in previous crop time or it should be less infested with diseases or pests.
    - It is appropriate to cultivate in block of land than in small plots of land for seed production. It is better not to cultivate seed for seed production for two years if loose smut disease appears in the land.
    - It is necessary to maintain spacing between the plots of seeds grown and seeds grown for cereal food of different variety for seeds as per certified standard. To maintain spacing, a small plot between the cultivated plots should be left empty or barren.
    - If the seed crop is self-pollinated crop; there will be less problem of external pollination. However; the isolation distance should be maintained in order to avoid transfer of diseases.
    - Three meters of isolation distance between the different varieties of seed crop cultivated for seeds or for different purposes is required. If any variety of seed is infested by loose smut disease, then a spacing of about 150 meters is necessary between the plots.

Soil test

Soil test refers to one or more of wide variety of soil analysis which is conducted for one of several possible reasons. Soil test shows analysis of soil sample which determines nutrient content, composition, and other characteristics like acidity or pH level. A soil test determines fertility, or expected growth potential of the soil which shows nutrient deficiencies, potential toxicities from excessive fertility and inhibitions from presence of non-essential trace minerals.

Soil testing is performed by commercial labs that offers variety of tests, targeting groups of compounds and minerals. It involves collecting of soil samples, preparation for analysis, chemical or physical analysis, interpretation of analysis results, and finally making fertilizer and lime recommendations for crops. The laboratory tests often checks for plant nutrients in three categories such as:

- Major nutrients: nitrogen (N), phosphorus (P), and potassium (K)
- Secondary nutrients: sulfur, calcium, magnesium
- Minor nutrients: iron, manganese, copper, zinc, boron, molybdenum, chlorine

Soil testing is an important diagnostic tool which determines nutrient needs of plants and for environmental assessments. It facilitates fertilizer composition and dosage selection for land employed in both agricultural and horticultural industries. Soil testing can also be used to identify application rates of waste materials containing nutrients or other elements that could harm the environment.

Soil testing is the best method available for determining how much nutrient is needed for optimum production of most crops. For certain crops, use of plant analysis serves as preferred method of determining crop nutrient needs. Making nutrient-application decisions without soil testing or plant analysis can be a risky endeavor because this will probably result in reduced profitability from either under or over-applying nutrients; in addition, over-application of nutrients can result in reduced water quality. It is important to follow recommended procedures for taking soil and plant samples.

3.1.2 Preparing field for seeds

- When establishing new plantation, certain actions need to be implemented to ensure the long term success of seed plantation which involves initially land preparation that should be done before transplanting of plant material.
- The idea of land preparation is to provide necessary soil conditions which enhance the successful establishment of young offshoots or tissue culture plants.
- Once a suitable area for seed plantation is selected, then planning operation gets finalised and actual preparation can be started.
- Such activities are divided to structure and to speed up implementation process for planting at most suitable time, as per specific regional climatic conditions.

Land preparation involves:

1 Mechanical field preparation

- The mechanical/initial soil preparation relates to preparation of field for further detailed preparation which includes irrigation system installation, hole preparation, etc which involves:
    - debushing/bush clearing;
    - removal of stones and rocks;
    - ripping;
    - levelling of soil

2 Irrigation system installation

- The type of irrigation system to be used will be determined by availability of water, topographical and soil conditions. When initial soil preparation is completed, the installation of required irrigation system will be implemented as per prescribed design.

3 Soil improvement

- The scheduling of soil improvement programme depends on the grower, as certain applications could be combined with initial actions of soil preparation.
- Due to long waiting period, planting to first production, it is a trend to establish plantations on new soils, with exception of areas where plantation is used for intercropping.

4 Hole preparation

- The actual digging of hole is done before the planting takes place, but it is seen that this is not the final preparation for planting operation itself as it is the point where required inputs like gypsum and organic materials are worked in soil.
- For this, a hole of 1 m^3 be prepared and that the soil from the hole be mixed with the organic material and gypsum and soil mix is put back in the hole where site is marked for positioning of small plants.

3.2 Selection of suitable area for quality seed production

3.2.1 Quality seed production

The quality of seeds is considered as an important factor for increasing yield. The use of quality seeds helps greatly in higher production per unit area to attain food security of the country. Seed quality is determined by number of genetic and physiological characteristics. The genetic component involves differences between two or more genetic lines, while differences between seed lots of a single genetic line comprise the physiological component.

The genetic factors that can influence quality include:

- Genetic make-up
- Seed size
- Bulk density

The physical or environmental characteristics include:

- Injury during planting and establishment
- Growing conditions during seed development
- Nutrition of the mother plant
- Physical damage during production or storage by machine or pest
- Moisture and temperature during storage
- Age or maturity of seed

High quality seeds are the result of good production practices, which include:

- Proper maintenance of genetic purity
- Good growing conditions
- Proper timing and methods of harvesting
- Appropriate processing during threshing, cleaning and drying
- Appropriate seed storage and seed distribution systems.

3.2.2 Land selection for quality seed production

- The land/field should meet seed production requirements.
- For certification purposes, field/land requirements of enterprise should meet selection requirements.
- The land must be ecologically suitable.
- The choice of right site is a prerequisite for proper seedling establishment which reduces seedling damage and stuck during transplanting.
- The land selected for seed production must be fertile, preferably light textured, with adequate irrigation and proper drainage system.
- The field should be free from weeds and volunteer plants from previous paddy crop.
- The field should not be infested with serious pests and diseases.
- Field should be well isolated as per seed isolation and previous cropping requirements.
- Field must be properly rogued, weeded and fertilized for good grain filling and optimum yields.

3.3 Record keeping, documentation and estimation of input required, received, output produced, sold quantity, profit and loss

3.3.1 Records

Keeping track of what to buy and knowing happening on fields requires some records. Good field management requires good useful set of field records which does not ensure successful of fields, but success is unlikely without them.

Record-keeping is a task that has been started many times by eager farmers/growers, only to be dropped later. Reasons that complete, accurate records are not kept revolve around lack of time and motivation. Keeping records is productive work.

Good record-keeping system should be taken into account for seed growers as per needs, abilities and limitations. Seed growers should need to understand the purpose for keeping the records and when the purpose is known, it will help them in deciding which records are useful and need to be kept.

With accurate facts and figures is useful while borrowing money, seeking government support and completing tax returns. From reliable farm records, equity proportion of total assets owned can be updated to assist with future farm investment programs.

Field records can help seed growers in evaluating seed enterprise on mixed field while field advice will be more effective if adviser knows exactly what is happening on the field. Records and statements show what has occurred or is occurring on the field while budgets predict what might happen in the future.

Records provide seed grower with data, information and knowledge. There are four uses for field records such as:

- Use as service tool which covers income tax calculations, estate planning, business arrangement reconciliation and obtaining and managing credit.
- Providing data for financial analysis and diagnostic instruments like identifying strengths and weaknesses of business.
- Progress indicator where series of records are required to monitor work progress.
- Forward planning tool for keeping past records to project future cash flows for current and modified farm management practices.

3.3.2 Details of Record

Records of field comprises of:

Field diary

- It is a form of small notebook where seed grower should record key facts and figures about farm business and its day-to-day activities.
- Farmers /seed grower who uses diary finds important facts and figures which could be easily lost or forgotten are stored permanently in this for future reference.

Seed Crop records

It is useful to record what happens to each plot or seed crop each year that will includes:

- crop,
- fertilizer applications,
- agronomy (seeding, weeding, sprays),
- visual impression of crop,
- harvest interval of forage crops
- seed crop yield

Unit costs

It covers all main field inputs like:

- fertilizers,
- fuel,
- irrigation water,
- concentrates and their ingredients,
- purchased forages,
- stock purchases

The details are required for routine book keeping and in monitoring seasonal changes and will help in planning future purchases.

Unit returns

- It includes returns from all field outputs such as sale of crop, manure etc.

Crop stock accounting

- The main purpose of seed crop stock accounts is to monitor net losses and gains in income, and to distinguish the increases and decreases due to changes in market value.
- When quantifying total stock value, seed grower has to take into account both in change size and change in unit of the crop.

Plant and improvement records

It covers records related to:

- date and cost of purchase or installation,
- annual depreciation,
- insurance and registration,
- fuel use,
- hours use,
- repairs and maintenance

Key financial records

- It covers detail related to interest and principal repayment schedules to plan repayments.

Other Field Charges

It include records related to:

- rates and government charges,
- telephone and office upkeep charges,
- labourer's wages,
- capital investments,
- unusual weather events,
- important meetings,
- credit repayments,
- changes in valuation of capital

Personal expenses

- It includes expenses related to personal input wages which includes valuation of any farm produce used for personal consumption.

Different records

- It covers cost related to crops, fertilizer, seeds, spray, labour, water, machinery, harvesting, processing, transport and selling.
- In this, wages are broken down into different farm activities.

3.3.3 Importance of Records

- They are management tool which allows seed growers to measure how efficiently they are using resources and to determine whether or not they are having income.
- The records are essential for planning and decision making
- The records are used in obtaining credit
- The records are used for income tax

3.3.4 Field Accounting

It is the process of measuring and recording all field resources and all field transactions with financial consequences in systematic way. The conduct of field accounting takes large part of seed grower/farmer's time and effort which gets compensated in account of following reasons:

- It permits farmer/seed grower to find out the size of income which is derived from the farm.
- It helps in knowing the total value of field business and to know which part is actually owned by him/her and which by others.
- It will help in detecting loss or theft of cash or stock.
- It provides necessary data for correct income tax assessment.
- It helps in claiming expenses.
- It helps in keeping complete information as farmers/seed grower dislike paper work since they are busy with their field work.

3.3.5 Field Financial Records

Field transactions can be recorded when they occur or at the time of cash transactions. Since farmers or seed growers are involved in lot of cash transition, so they should use accrual method of accounting, which will help them in tracking of their money, inventory, sales etc. For example, when recording sales of seed for specific time period, all the expenses needed to generate that seed are relevant when calculating the cost of production.

Every physical item on field can be placed into following categories:

- Assets: It is required to generate income which can be either owned/purchased on credit that can be combination of physical and monetary values, where physical quantity is multiplied by unit price to have its value.
- Liabilities: It relates to items on credit.
- Equity: It serves as an asset with less liabilities.

- Revenue: It is source of income that is generated from the business.
- Expenses: It involves the costs what is required for a business to generate revenue.

Annual financial accounting comprises of:

- Opening balance sheet: record all assets and liabilities at beginning of financial period.
- Closing balance sheet: record all assets and liabilities at end of financial period.
- Cash flow: record all cash inflows and outflows in business.
- Income statement: calculate wheat enterprise profit/loss before or after income tax.

3.4 Soil sampling for soil testing

3.4.1 Soil sampling

Soil sampling is done for checking of soil pH, phosphorus, potassium and micronutrients which should be done after every three to four years as these values do not change rapidly even with fertilization. Soil samples for soil nitrate should be taken 1-3 months before planting each wheat crop. In most of dryland or rainfed wheat production areas, wheat root will have depth of 5-foot with full profile of soil water.

Effective and profitable fertilizer recommendations require accurate sample of the field. It is very important that the soil sample should accurately represent the field to be fertilized. Previous crop history, previous lime and fertilizer rates and soil characteristics will cause variance in results if not accounted while sampling.

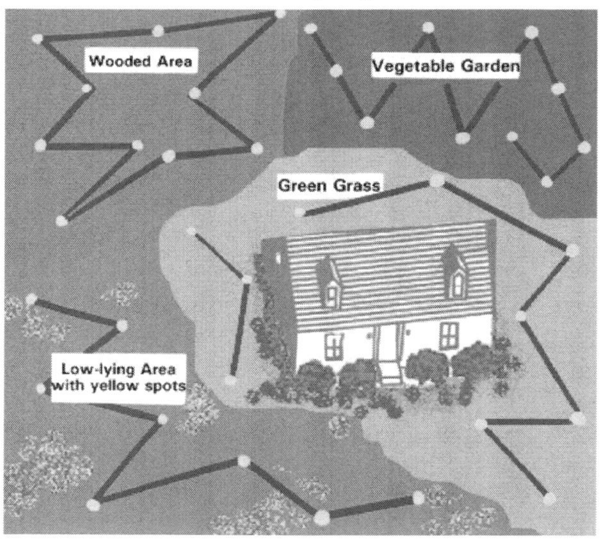

- Soil sampling is most challengeable work, as few grams of soil sample represents for a given area, hence soil samples needs to be taken in such a manner that the collected sample should reflect the true fertility of soil for any part of the land area.
- If the field is levelled and soil appears to be uniform, only one composite sample if taken properly could be enough for an area of 4-5 ha.

- Variation in slope, colour, texture, crop growth, unusual spots and management practice should be taken into account and separate sets of composite samples must be collected from each such area.
- Recently fertilized plots, bunds, channels, marshy tracts and area near trees, wells, cattle dung and compost piles or other non-representative locations must be avoided during the sampling.

3.4.2 Sampling Tools:

There are many tools that are used for soil sampling which includes:

- spade/khurpi/proble

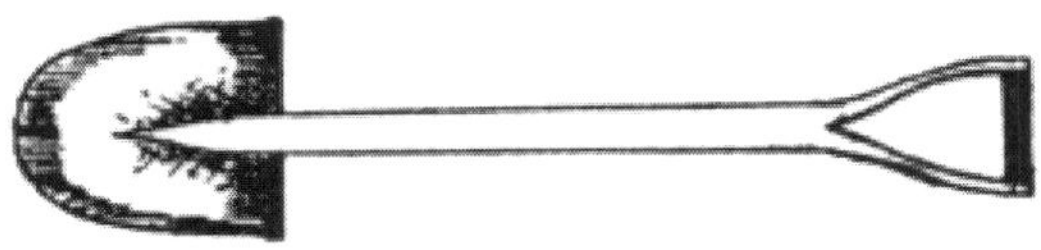

- auger (tube and screw type),

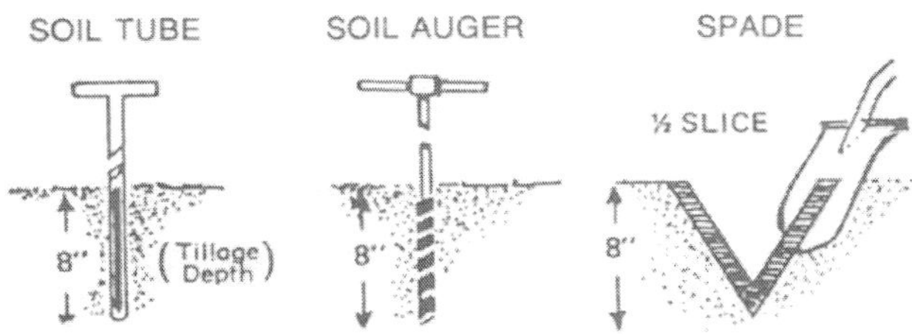

- plastic bucket,

- plastic bag,
- scale
- water proof marker

In case of soil sampling is done for soft and moist soil, the spade or khurpi is quite satisfactory while screw type auger is used for hard/dry land.

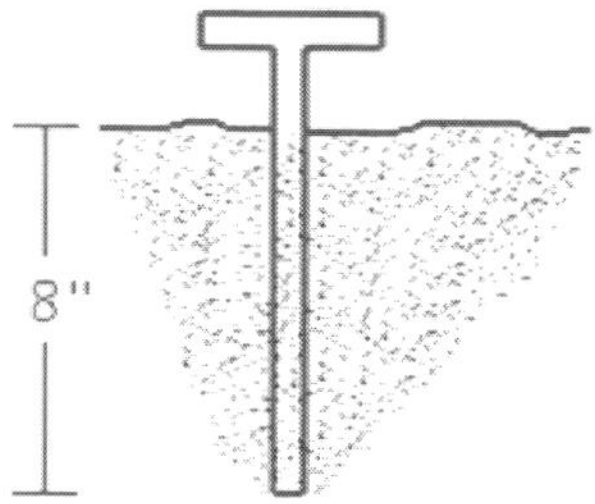

Soil probe/spade is the best tool for obtaining soil samples. For getting soil sample:

- ❖ Push the probe/spade to 8 inch depth and remove the core.
- ❖ Take a slice of soil about 1/2 inch thick, with the spade.
- ❖ Using a trowel, cut a 1/2 core lengthwise from the center of the slice while it is still on the spade. Put the core in a bucket to be mixed with the 15 other cores from the designated sampling area.

3.4.3 Selecting soil sampling area

- ❖ Sampling of soil is done by visual survey of the field where variation related to slope, colour, texture, management and cropping pattern are observed.
- ❖ The land selected should be demarcated into uniform portions so that each demarcated portion should be sampled separately.

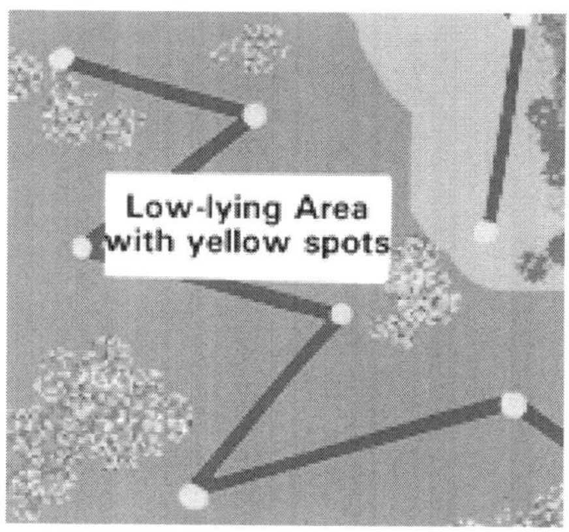

- ❖ If all conditions are same, then one field is treated as single sampling unit.
- ❖ The land selected should not be more than 1 to 2 hectares, and should be an area where a farmer is willing to give extra attention.

3.4.4 Sampling Depth

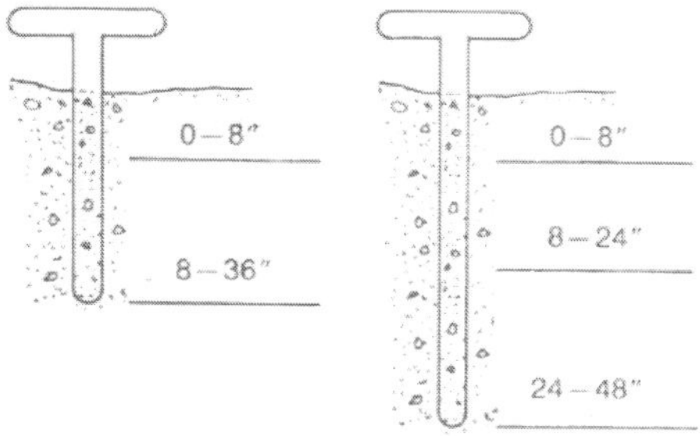

- ❖ Surface soil samples from fields should be taken to a depth of 8".
- ❖ Subsoil samples should be taken from 8" to 36" for residual nitrate.
- ❖ For better sampling, take sample from 8-24" and 24-48".

3.4.5 Sampling procedure

- ❖ If fields area is large, then select the part of field and divide into respective field area. If soil areas in field is different in appearance (slope, drainage, color or texture), then each soil area should be sampled separately.

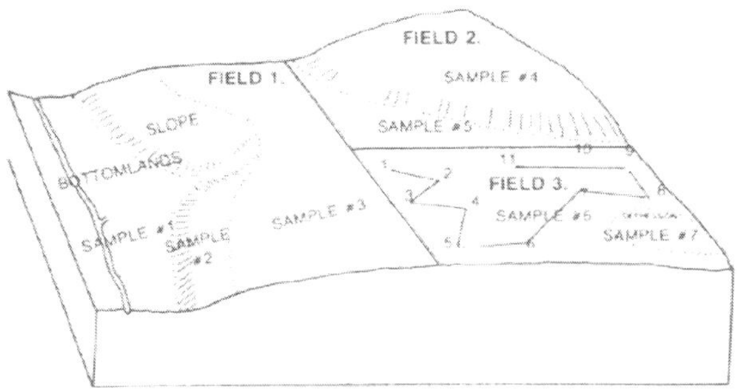

- ❖ Clear the area for sampling by removing or scrapping away stones, rubbish or trash from surface to expose the soil before sampling.
- ❖ Once the land is demarcated, mark the portion of land in zig-zag manner by different sampling unit boundaries.

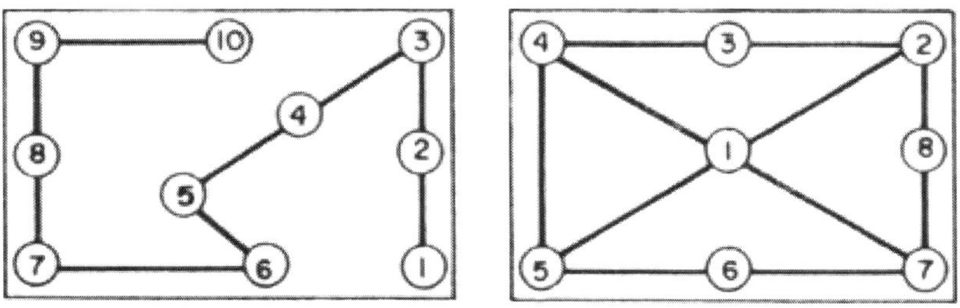

❖ On the plain with number of samples and manner of composite, enter the sampling and mark each unit with different letters/numbers.

❖ Now take a slice of the plough layer cut at intervals of 15 to 20 steps or as per area to be covered.

❖ Now remove away surface liter and mark the sample area from the soil sample to be taken.

❖ Using spade or trowel, get the soil sample at specified depth, i e 2.5 cm thick and 5 cm wide.

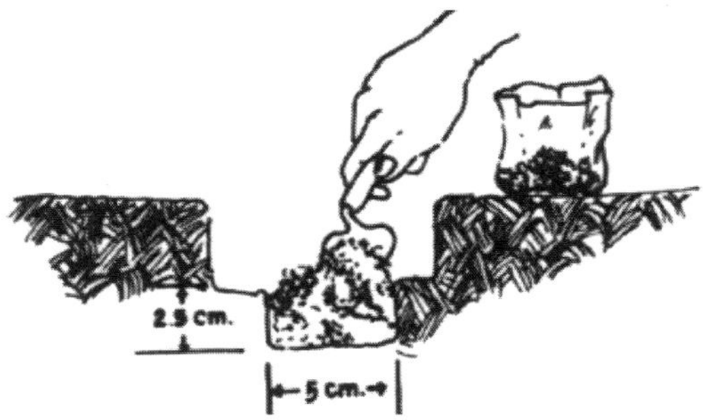

❖ Mark V shaped cut to a depth of 15 cm in the sampling spot with the help of spade to remove 1 to 2 cm slice of soil.

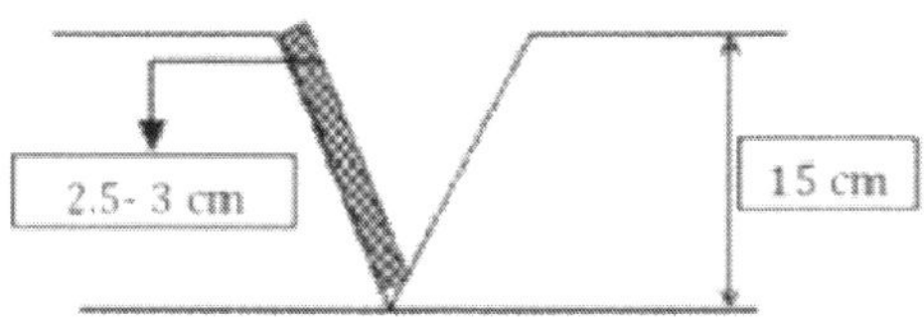

❖ Now collect the sample on the blade of the spade and put in a clean bucket.

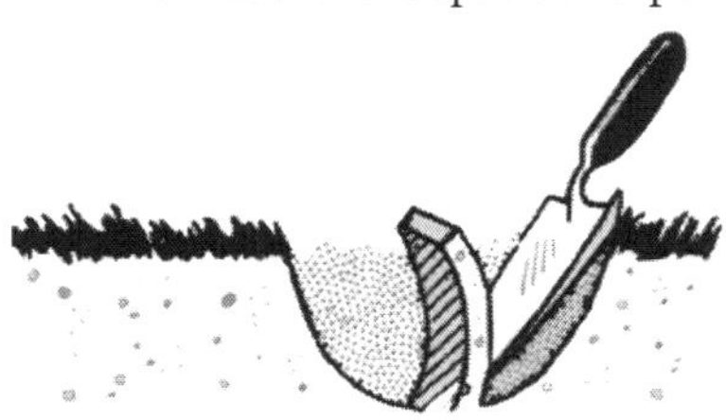

❖ Collect soil sample and place it in a clean bucket or tray.

❖ Now take samples from 10 to 20 spots as marked for one composite sample as per size of the field.

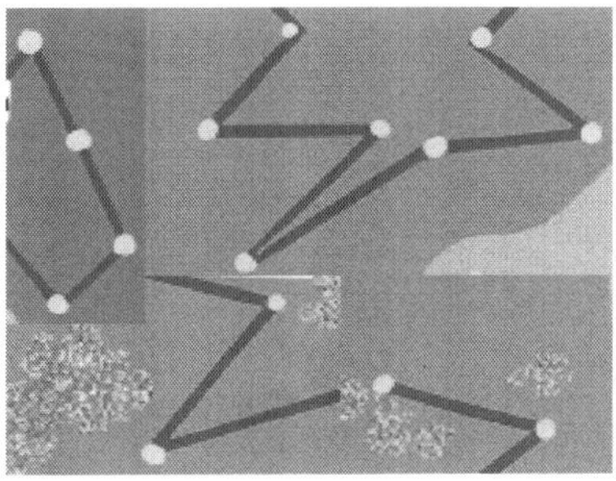

❖ Collect the different soil samples with the help of augur from the plough depth and keep them in separate glass jar.

❖ The glass jar should be marked with any reference, so that the samples can be easily identified.

3.4.6 Preparation of soil samples

Preparing the collected soil sample for testing in laboratory involves the following steps:

❖ Soil samples which are collected from different units usually contain moist, aggregated state, etc that is not suitable for laboratory analysis.

❖ The soil sample is collected in jar/bucket should be free from organic debris, rocks and trash and needs to be mixed properly.

❖ Pour the soil sample from the bucket/jar on a piece of clean paper or cloth and mix it thoroughly.

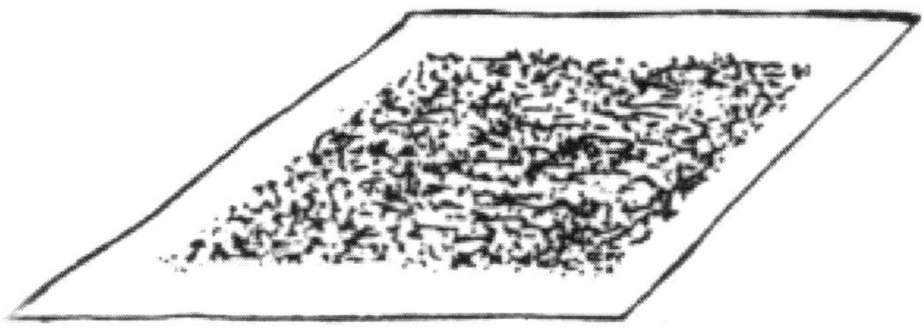

❖ The soil sample spread on a plastic or on thick paper should be dried and then grinded to avoid contamination.

❖ Field moist samples before drying, should pass through a 6 mm sieve by rubbing with fingers.

❖ After sieving, samples are air-dried (25-35° C; relative humidity 20-60%) and stored.

❖ Samples should not be exposed directly with the sunlight, but should be kept under the shade for drying.

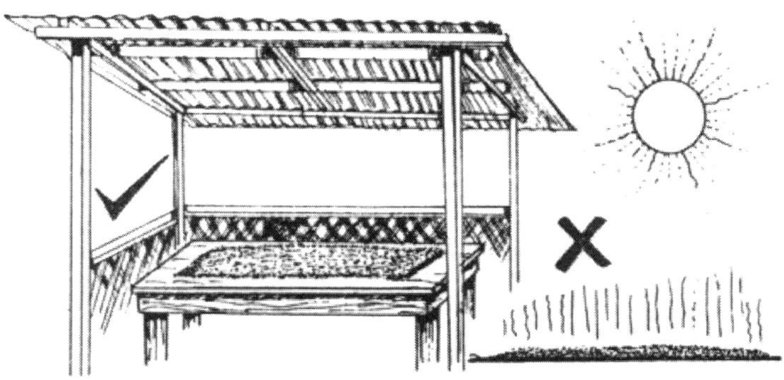

- To speed up the drying process, break the samples into different compartments.

- Keep the dried soil sample in a clean plastic bag free from any contamination.
- Write details of the sample in the information sheet and put it in the plastic bag.
- Tie the mouth of the bag carefully.

3.4.7 Submission of soil samples for analysis

Soil test determines soil nutrient capacity by mixing soil during the analysis with strong extracting solution. The soil reacts with the extracting solution, thereby releasing some nutrients. As soil supplies most of the mineral nutrition for higher plants through plant's root system, the extracted nutrient concentration is evaluated as per research that relates plant utilization to soil nutrient concentrations.

Soil testing laboratory is mainly used for testing soil samples which provides an extension advisory to the farmers by giving soil testing reports with advice and recommendations on soil condition and suitable crop that can be cultivated on particular soil.

Soil sample collected from different field units are supplied to soil testing laboratories. Standard or routine soil tests vary from laboratory to laboratory which includes analysis about:

- soil texture;
- electrical conductivity
- soil pH;
- available phosphorus (P),
- potassium (K),
- calcium (Ca),
- magnesium (Mg);
- sodium (Na);
- cation exchange capacity
- OM content

As part of soil analysis, laboratory usually supplies some interpretation, which covers indication of whether individual soil tests are low, medium or high. The laboratory provides fertilizer recommendations based on the analysis, as these recommendations are plant and soil specific.

Soil analyses are relates to potential nutrient uptake, supplementation of plant nutrients through fertilisation and target yield. Soil testing laboratory serves as production line raw materials where soil samples are received at one end and finished products report emerges at other end. The production line has series of operation that should be done in logical sequence with operations simultaneously.

3.4.8 Soil Analysis

1. Organic Matter

Percent organic matter is a measurement of the amount of plant and animal residue in the soil. The color of the soil is usually closely related to its organic matter content, with darker soils being higher in organic matter

2. Phosphorus

Shows three types of phosphorus tests; P1 (weak Bray), P2 (strong Bray) and Bicarbonate P (sodium bicarbonate).

3. Potassium

It measures the available potassium in the sample where its optimum level varies with crop, yield, soil type, soil physical condition, and other soil related factors.

4. Magnesium and Calcium

It shows the levels of calcium and magnesium in the soil that are affected mainly by soil type, drainage, liming and cropping practices.

5. Sodium

It shows the sodium level in soil sample

6. Soil pH

This shows soil pH measures, active soil acidity or alkalinity in soil sample.

7. Buffer Index

It shows an index value for determining the amount of lime to apply on acid soils with pH less than 7.0.

8. Cation Exchange Capacity

It measures the soil's ability to hold nutrients like potassium, magnesium and calcium as well as other positively charged ions as sodium and hydrogen in soil sample.

9. Percent Base Saturation

It shows part of Cation Exchange Capacity occupied by given cation in soil sample.

10. Nitrate-Nitrogen

The soil test measures nitrate-nitrogen (NO3-N) which is water soluble and readily available for plant uptake.

11. Sulfur

The soil test measures everal forms of sulfur that can be readily available.

12. Zinc

It shows extraction of the zinc in soil sample.

13. Manganese

It shows detail about manganese extraction reading using DTPA extraction process.

14. Iron

It shows details about iron extraction using the DTPA extraction process in the sample.

15. Copper

It shows detail about copper extraction using DTPA extraction process in the sample.

16. Boron

It shows reading of boron extraction from the soil using DTPA/Sorbitol where adequate levels shows from 1 to 1.5 ppm.

17. Excess Lime Rate

It shows detail about visual rating of free lime where soil with high amounts of free lime have problems with availability of major and minor elements to the plant.

3.5 Identification of equipments used in seed production

Seed Scalper

- The equipment removes large inert matter from seeds.
- When this equipment contains single sieve it is known as scalpers while two sieves equipment is known as rough cleaners.
- This unit has vibrating or rotating screen or sieve with perforation to allow rough seed to pass through.

Debearders

- The equipment has horizontal beater with arms rotating in steel drum.
- In this, when seeds pass through the equipment, it does an action of rubbing of seeds and clipping of oats seeds, debeard barley, thresh white cap in wheat, remove awns and beards, de hull some grass seeds and polish the seed.

Huller - scarifier

- The equipment has two rubber faced rough surfaces that rub the seeds.
- With this equipment, dehulling and scarifier can be done simultaneously or separately.

❖ The equipment has many items, seed – rotating disc, centrifugal force – thrown – huller, suction chamber.

Air-Screen cleaner and grader

❖ The air-screen equipment is a cleaner used in seed processing plants.

❖ In this, all seeds are cleaned by air-screen cleaner before doing particular specifications.

❖ The equipment size varies from small, two-screen farm models to large industrial cleaners with 7-8 screens.

❖ In such machines, separations are done based on differences in single physical characteristic.

Specific gravity separator

❖ In this equipment, mixture of seeds is fed on lower end of sloping perforated table where air is forced up by porous deck surface and bed of seeds by fan, which stratifies seeds in layers as per density with light seeds and particles of inert matter at top and heaviest at bottom.

Spiral separator

❖ The separator, which classifies seed as per its shape and rolling ability, has sheet metal strips fitted around central axis in form of spiral.

❖ In this equipment, unit resembles an open screw conveyor standing in vertical position where seed is introduced at top of inner spiral.

❖ Round seeds roll faster down the incline which are flat or irregular shaped, that tend to slide or tumble.

3.6 Understand and perform different methods of tillage, seed sowing techniques and other field management practices

3.6.1 Tillage

❖ Tillage is an agricultural preparation of soil using mechanical agitation such as digging, stirring and overturning.

❖ The human-powered tilling methods with hand tools include shovelling, picking, mattock work, hoeing, and raking.

❖ Tillage has been used for millennia to prepare soil before sowing many of annual grain crops which involves applying power to break up and rearranging complete topsoil structure.

❖ The main aim of tillage is to destroy weeds and pests and also important for incorporating, redistributing or releasing nutrients and making soil texture suitable for seed sowing, seed germination and easy penetration of seedling roots.

❖ Tillage operation and cultivation in farming season comprises of plowing, preparing seed bed, leveling, cultivating, covering seed, making irrigation furrow, and application of fertilizer.

- Machines used for tillage operation consist of different kinds of plowshare, disk, harrow, cultivator, leveler, planter and furrow opener that gets attached to tractor and are used in farm to do various operations.

Need of till

- Soil disturbance before or during seed placement ensures that sowing depth is optimal and consistent.
- To ensure that seedling coleoptile can penetrate its covering of soil and roots rapidly to explore soil to its depth.
- Seeds are sown at wrong depth or at irregular depth due to poorly prepared cloddy soils or inadequate seeding equipment which gives poor yields.
- The aim of soil disturbance is to set back or kill any weeds and pathogens that will restrain growth of developing seedlings.

Effects of tillage

- It loosens and aerates top layer of soil/horizon A, which facilitates planting the crop.
- It helps in mixing harvest residue, organic matter (humus) and nutrients evenly in the soil.
- It mechanically destroys weeds.
- It dries soil before seeding.
- It helps in exposing soil crumble over winter through frosting and defrosting, that prepares smooth surface for spring planting.

3.6.2 Methods of tillage

There are three methods:

(1) Conventional tillage as maximum practice,

(2) Reduced tillage as a minimum practice

(3) No-till

3.6.3 Techniques of seed sowing

Planting method plays an important role in the placement of seed at proper depth, which ultimately affects crop growth. The selection of suitable sowing method plays an important role in the placement of seed at proper depth, which ensures better emergence and subsequent crop growth. Wheat is planted with different sowing methods depending upon the available soil water, time of planting, amount of residue in the field and availability of planting machine.

Broadcasting:

- The seeds are broadcast and mixed in the soil using harrows.
- The uniform seed distribution cannot be achieved leading to irregular plant stand and poor germination.

- ❖ If broadcasted seeds are not covered properly, they are picked up by birds resulting in wastage of seeds.

Behind Local Plough

- ❖ The seeds are dibbled by hand into the furrows that are opened with the plough known as kera (hand) method.
- ❖ On dropping through Pora or Nai or Hazara, a special attachment with local plough is called Pora method.
- ❖ The seeds are dropped at a depth of 5 to 6 cm.
- ❖ The germination is satisfactory under this method of sowing.

Drilling:

- ❖ Before sowing seeds by drilling, seed-bed should be thoroughly prepared.
- ❖ The seed bed should be well leveled and free from clods and weeds.
- ❖ The seeds are sown by seed drilling at uniform depth.
- ❖ It results in uniform germination and regular stand.

Dibbling:

- ❖ In case of limited seed availability, dibbling method is used for sowing.
- ❖ Dibblers have a wooden or iron frame with pegs.
- ❖ The frame is pressed in the field and lifted and then one or two seeds are dropped by hand in each of the hole.
- ❖ It is a time consuming process.

Zero Tillage Technique:

- ❖ It is a useful method under rice-wheat cropping system.
- ❖ Once the rice crop is harvested, field requires at least 6-8 tillage operations to make field ready for sowing of wheat.
- ❖ A zero-till-seed drill machine is used for direct sowing of wheat in rice field without field preparation which helps in advancing of sowing of wheat.
- ❖ Under zero tillage, there should be proper moisture in the field at the time of sowing.
- ❖ The rice crop should be harvested close to the ground.
- ❖ The field should be free from weeds.
- ❖ Sowing depth of about 5 to 6 cm should be maintained.
- ❖ Light plank may be used behind the machine.

3.7 Understand and perform rouging, weeding and field inspections at various growth stages of seed crop

3.7.1 Rouging

- It is removal of plants which are off-type that is phenotypically different from plants of variety under production.
- It is important aspect of seed production which is needed to save out-crossing and mechanical mixtures.
- The off-type plants are regularly removed from field by uprooting or cutting at ground level.
- It involves in removal of mixtures of plants.
- It involves in any seed production plot which is important to prevent pollen from off-type plants causing irreparable damage to crop by cross-pollination.

Features:

- It is selective removal of undesirable plants from seed crop based on visual field inspection, to improve qualities such as genetic purity, disease free attributes of seed to be harvested.
- It involves removal of noxious weeds that are liable to multiply with seed crop, which affects future generations.
- At all stages of crop in field is a need to maintain variety purity as it was at time of release/notification.
- As rogue plants are not distinguishable before flowering, so rouging should be done before blooming starts.
- In this, doubtful plants also need to be rouged.
- Rogued plants remove from field soon after rouging and destroying as it survives for few days which spreads the pollen.

Rouging consists of removal of:

- Off types
- Volunteer plants
- Pollen shedders in female
- Plants of noxious weeds and other crops
- Diseased plants affected by seed borne diseases growing in seed plot
- Tassels from plants in female rows of seed production of single and double hybrids in crop.

Phases:

Rouging is carried in three phases:

- Pre Flowering

- During Flowering
- Post flowering or before harvesting

3.7.2 Weeds

- Weed is a plant considered undesirable in a particular situation.
- It includes plants unwanted in human-controlled settings like farm fields, gardens, lawns and parks.
- It has no botanical significance as a plant that is a weed in one context is not a weed when growing in a situation where it is in fact wanted, and where one species of plant is a valuable crop plant, another species in the same genus might be a serious weed, such as a wild bramble growing among cultivated loganberries.
- Volunteer crops are regarded as weeds in a subsequent crop.
- Many plants that people widely regard as weeds also are intentionally grown in gardens and other cultivated settings, in which case they are sometimes called beneficial weeds.
- Weed applies to any plant that grows or reproduces aggressively, or is invasive outside its native habitat.

3.7.3 Field inspection

- It is an inspection of standing seed crops which is required in order to verify the conformity of seed crops as per prescribed certification standards.
- The main objective in conducting field inspection is to verify factors which can cause irreversible damage to genetic purity or seed health.
- Field inspection needs to be carried out in recommended/prescribed crop stages.
- It involves verification of all factors thatt affects seed quality in field which cannot be possible in single inspection.
- In this, all factors in field may not be apparent or capable of affecting seed quality at certain stage of crop growth.
- It covers all important stages of crop growth where number of inspections and stages of crop growth varies from crop to crop as per duration, nature of pollination, susceptibility to contamination, disease susceptible stages, nature of contaminating agents, seed production procedures etc.

Objectives

The objective of field inspection is to check:

- seed origin and identity of variety.
- information on cropping history of seed field.
- crop and cultivation conditions.
- isolation distance among the field crops.
- freedom from impurities like other crop plants and weeds plants.
- freedom from seed-borne disease.

3.8 Seed production technology of various field crops

3.8.1 Seed Production Protocol

The seed production process/protocol starts with proper identification of seeds bearing:

- clear scientific name
- origin of seed distribution
- description about its production

The process/protocol for production of planting materials includes seeds from clean planting material source which covers:

- Field facilities and equipment
- Source of material
- Field requirement
- Field inspection
- Agronomic practices like isolation, rotation and negative selection
- Harvesting and handling
- Post-harvest treatments
- Storage and transport
- Quality standards for supplied product

3.8.2 Factors responsible for seed production

The quality of seeds is considered as an important factor for increasing yield. The use of quality seeds helps greatly in higher production per unit area to attain food security of the country. Seed quality is determined by number of genetic and physiological characteristics. The genetic component involves differences between two or more genetic lines, while differences between seed lots of a single genetic line comprise the physiological component.

The genetic factors that can influence quality include:

- Genetic make-up
- Seed size
- Bulk density

The physical or environmental characteristics include:

- Injury during planting and establishment
- Growing conditions during seed development
- Nutrition of the mother plant
- Physical damage during production or storage by machine or pest
- Moisture and temperature during storage
- Age or maturity of seed

Good quality seeds are result of good production practices, which include:

- ❖ Proper maintenance of genetic purity
- ❖ Good growing conditions
- ❖ Proper timing and methods of harvesting
- ❖ Appropriate processing during threshing, cleaning and drying
- ❖ Appropriate seed storage and seed distribution systems.

Unit 4

Seed Health Management

4.1 Perform management practices; nutrients/manure/weeds/irrigation/ growth regulators/pesticide spraying/ fertilizers doses/pollination practices/other management in seed production

4.1.1 Weeds and its types

Introduction

- Weed is a plant considered undesirable in a particular situation.
- It includes plants unwanted in human-controlled settings like farm fields, gardens, lawns and parks.
- It has no botanical significance as a plant that is a weed in one context is not a weed when growing in a situation where it is in fact wanted, and where one species of plant is a valuable crop plant, another species in the same genus might be a serious weed, such as a wild bramble growing among cultivated loganberries.
- Volunteer crops are regarded as weeds in a subsequent crop.
- Many plants that people widely regard as weeds also are intentionally grown in gardens and other cultivated settings, in which case they are sometimes called beneficial weeds.
- Weed applies to any plant that grows or reproduces aggressively, or is invasive outside its native habitat.

Features

- A number of weeds like dandelion and lamb's quarter, are edible with leaves or roots used for food or herbal medicine.
- It prevent pest insects from finding a crop as their presence disrupts the incidence of positive cues which pests use to locate their food.
- Weeds act as a living mulch and provides ground cover to reduces moisture loss and in preventing erosion.
- It improves soil fertility by bringing up the nutrients like calcium and nitrogen from deep in soil with tap root and clover hosts nitrogen-fixing bacteria in its roots, fertilizing the soil directly.

- The dandelion species breaks up the hardpan in overly cultivated fields and help crops to grow deeper root systems.
- Many garden flowers originates as weeds in cultivated fields and have selectively bred for garden worthy flowers or foliage.

Weed Control Methods

Controlling weeds in crop production involves wide range of techniques which includes:

Preventative Weed Control

- It refers to control method that aims to prevent weeds from being established in a cultivated crop, a pasture, or a greenhouse.
- Examples include use of certified weed free seed, transporting weed free hay, making sure farm equipment is cleaned before moving from one location to another, and screening irrigation water to prevent weed seeds from traveling along irrigation ditches.

Cultural

- It refers to a technique that involves maintaining field conditions such that weeds are less likely to become established and/or increase in number.
- Examples of cultural weed control would be crop rotation, avoiding overgrazing of pastures or rangeland, using well-adapted competitive forage species, and maintaining good soil fertility.

Mechanical

- It refers to technique that involves the use of farm equipment to control weeds.
- The two mechanical control techniques most often used are tillage and mowing.

Biological

- It refers to technique that involves the use of natural enemies of weed plants to control the germination of weed seeds or the spread of established plants.
- It is rapidly expanding area of weed control with many examples which includes sheep to control tansy ragwort or leafy spurge, cinnabar moth and the tansy flea beetle to control tansy ragwort, the chrysolira beetle to control St. John's Wort, and the use of goats to control brush on rangeland.

Chemical

- It refers to technique that involves the application of a chemical (herbicide) to weeds or soil to control the germination or growth of the weed species.
- It includes chemicals to control weeds which includes 2,4-DB; EPTC; bromoxynil; and paraquat.

4.1.2 Water need for Crop

- Water is one of the most important inputs essential for production of crops.
- Plants need water continuously during their life in huge quantities.
- It influences photosynthesis, respiration, absorption, translocation and utilization of mineral nutrients, and cell division apart from other processes.
- Water shortage and excess will affect the growth and development of a plant directly and, consequently, its yield and quality.
- Water affects the performance of crops not only directly but also indirectly by influencing the availability of other nutrients, the timing of cultural operations, etc.
- Water and other production inputs interact with one another.
- In proper combinations, crop yields can be boosted manifold under irrigated agriculture.
- It is a costly input when canals supply it.
- The constructing of dams and reservoirs, the conveying of water from storage points to the fields, the operating and the maintaining of canal systems involve huge expense.
- The misuse of water leads to problems of water logging, salt imbalance, etc. which render agricultural lands unproductive.
- Proper appreciation of relationship and economic utilization of water resources is required for maximum crop production.

Water requirement for crops

- It is essential to know the water requirement of a crop which is the total quantity of water required from its sowing time up to harvest.
- Naturally different crops may have different water requirements at different places of the same country, depending upon the climate, type of soil, method of cultivation, effective rain etc.

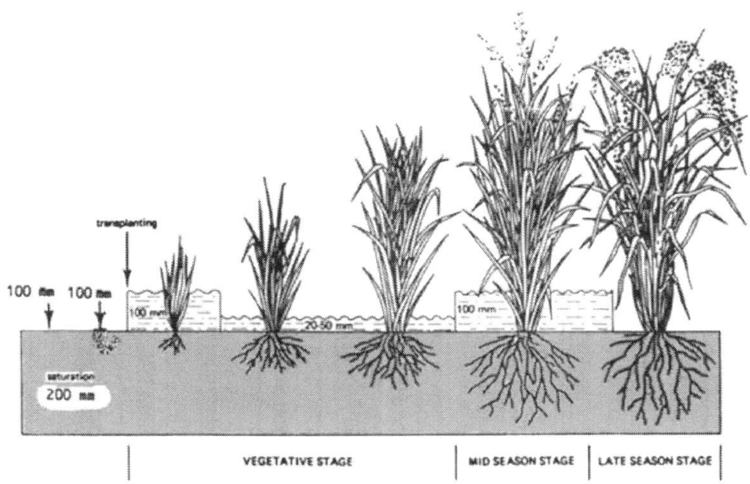

- The total water required for crop growth is not uniformly distributed over its entire life span known as crop period.
- The watering stops same time before harvesting and time duration from first irrigation at the time of sowing to last before harvest is known as base period.
- As water is essential for crop growth, it's requirement for each crop is definite which differs as per the season (Kharif, Rabi or Summer) or as per soil type (light, medium and heavy) but net requirement results same.

4.1.3 Irrigation system

The irrigation system consists of:

Intake structure

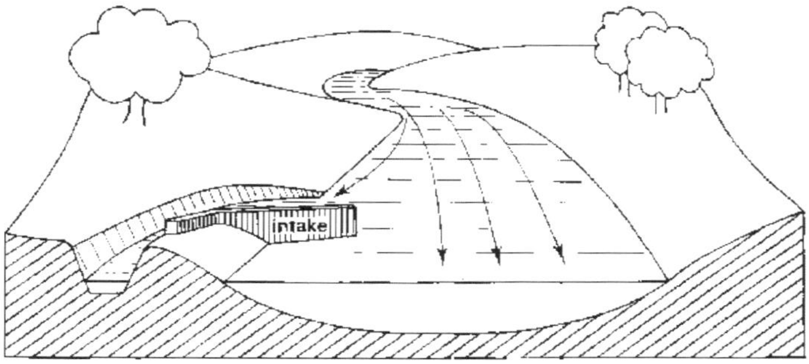

- The intake structure is built at the entry to the irrigation system.
- The purpose of intake structure is to direct water from the original source of supply (lake, river, reservoir etc.) into the irrigation system.

Pumping station

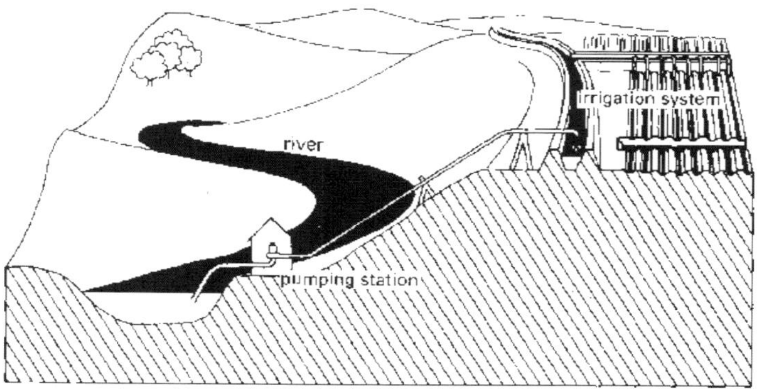

- Many times, irrigation water source lies below the level of the irrigated fields.
- The pump is used to supply water to the irrigation system.
- There are several types of pumps, but the most commonly used in irrigation is the centrifugal pump.

Conveyance system and distribution system

- ❖ The conveyance and distribution systems consist of canals transporting the water through the whole irrigation system.
- ❖ Canal structures are required for the control and measurement of the water flow.

Open canals

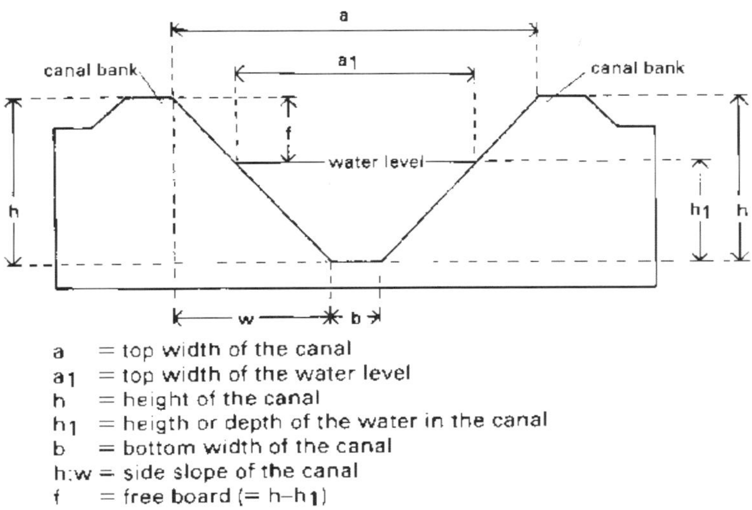

- ❖ An open canal, channel, or ditch, is an open waterway whose purpose is to carry water from one place to another.
- ❖ Channels and canals refer to main waterways supplying water to one or more farms.
- ❖ Field ditches have smaller dimensions and convey water from the farm entrance to the irrigated fields.
- ❖ The most commonly used canal cross-section in irrigation and drainage, is the trapezoidal cross-section.

Field application system

- The traditional methods of applying water to the field is by bringing the water from the source of supply, such as a well, to each plant with a bucket or a water-can.
- Surface irrigation is application of water to the fields at ground level in which, either the entire field is flooded or the water is directed into furrows or borders.
- Furrows are narrow ditches dug on the field between the rows of crops.
- The water runs along the furrow as it moves down the slope of the field.
- The water flows from the field ditch into the furrows by opening up the bank or dyke of the ditch or by means of syphons or spiles.

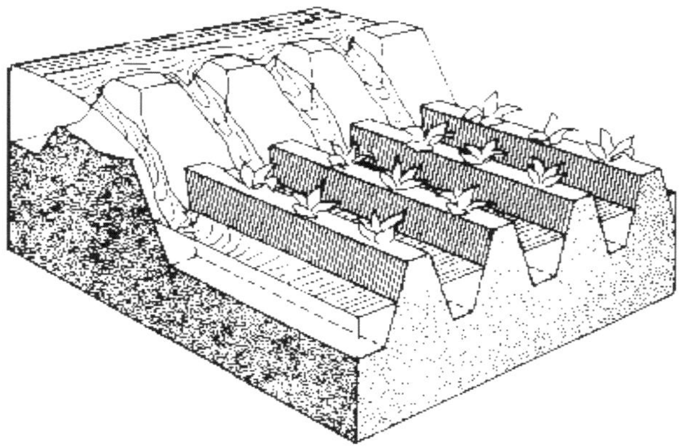

- Siphons are small curved pipes that deliver water over the ditch bank.

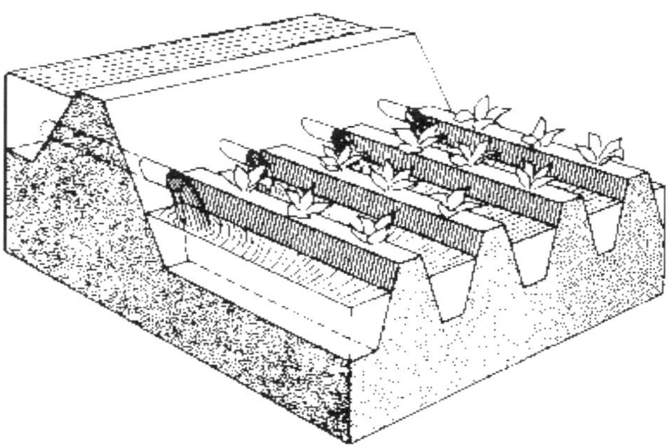

- Spiles are small pipes buried in the ditch bank.
- In border irrigation, the field to be irrigated is divided into strips by parallel dykes or border ridges.

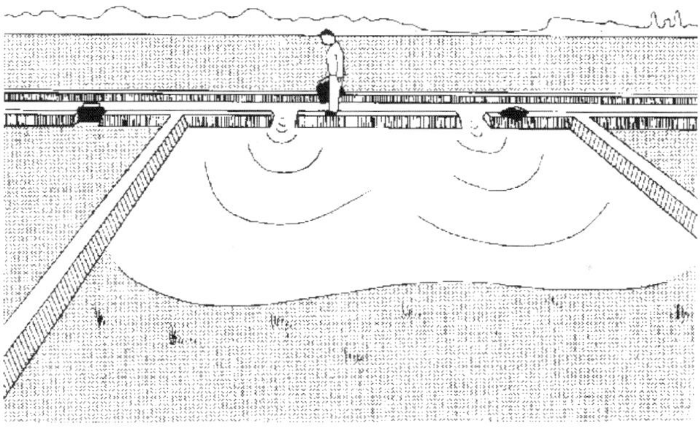

- In border irrigation, water is released from the field ditch on border through gate structures called outlets or by means of siphons or spiles.
- In basin irrigation, basins are horizontal, flat plots of land, surrounded by small dykes or bunds in which the banks prevent the water from flowing to the surrounding fields.

- Basin irrigation is commonly used for rice grown on flat lands or in terraces on hillsides.

- Trees can also be grown in basins, where one tree usually is located in the centre of a small basin.

Drainage system

- A drainage system is required to remove excess water from the irrigated land.
- The excess water may be e.g. waste water from irrigation or surface runoff from rainfall.
- Waste water can also be leakage or seepage water from the distribution system.
- Excess surface water is removed through shallow open drains.
- Excess groundwater is removed through deep open drains or underground pipes

4.1.4 Fertilizers

Organic

- ❖ It comprises of variety of plant-derived materials that ranges from fresh or dried plant material to animal manures and litters to agricultural by-products.
- ❖ The nutrient content of organic fertilizers varies greatly among source materials, and readily biodegradable materials make better nutrient sources.
- ❖ Nitrogen and phosphorus content is lower, often substantially lower, in organic fertilizers compared to chemical fertilizers.
- ❖ Moisture content is another factor that reduces or dilutes the nitrogen and phosphorus concentrations of organic fertilizers.
- ❖ It provides energy for biological activity.
- ❖ Many of the nutrients used by plants are held in organic matter until the organisms decompose the materials and release them for the plants' use.
- ❖ It attracts and holds plant nutrients in an available state, reducing the amount of nutrients lost through leaching.
- ❖ It improves soil structure, so that air reaches plant roots and the soil retains moisture.
- ❖ The organic matter and the organisms that feed on it are central to the nutrient cycle.

Determine right organic/inorganic fertilizers to be applied

- ❖ Inorganic fertilizers in combination with organic fertilizers help in improving soil health and sustain crop yield.
- ❖ For better crop production, both organic and inorganic plays an important role.
- ❖ Fertilizer use, though has been a prime contributing factor for the increased agricultural productivity in the areas, there are situations when they are not available in right time and adequate quantity despite high priced.
- ❖ After the soil test results, wheat cultivator can be able to judge the required supplemental fertilizer along with type of fertilizer to be used.

The factors which involves in selecting correct fertilizer include:

- ❖ physical properties
- ❖ chemical properties

- environmental stewardship
- economics

The good fertilizer selection involves:

- Solids, liquids or gases forms where each physical form has its own uses and limitations that provides the basis for selecting the best fertilizer for the field.
- Correct solubility, as it shows how readily the nutrients are dissolved in soil water and are taken by the plants.
- Required salt index as soluble salts at high concentrations in soil solution will injure/kill plants or prevent germination of seeds.
- Checking of acidity of some fertilizer materials.
- Correct particle size as it is important for producing stable, high-quality blended fertilizer.
- Comparison of material price per actual plant food by formula, (Price per ton of fertilizer)/(2000 x plant food content as decimal value) = per rupee of plant food

Apply fertilizers in right dosage, time and intervals of crop growth

- Timing of fertilizer application has a significant effect on crop yields.
- Proper timing of the fertilizer application increases yields, reduces nutrient losses, increases nutrient use efficiency and prevents damage to the environment.
- Applying fertilizers at wrong time might result in nutrient losses, waste of fertilizer and even damage to the crop.
- Plants need different nutrient rates and ratios at different growth stages.
- To make the nutrients available when plant needs them, fertilizers should be applied at the right timing.
- The optimum timing for fertilizer application is determined by Nutrient Uptake Pattern of the crop.
- For similar crop, each nutrient has an individual uptake pattern.
- The maximum rate of fertilizer that can be applied at one application depends on the salinity threshold that the crop can tolerate.
- Split fertilizer applications helps in avoiding salt damages to the crop and improving germination rate.
- Soil type affects the timing and frequency of fertilizer application where Cation Exchange Capacity and Soil Texture properties determine the frequency and timing of application.
- If nitrogen is applied too early before need of plant, then a required portion may get lost before the crop takes it up, so time nitrogen is in soil before the plant takes up should be minimized.

4.2 Understand need based application of pesticides/fungicides

4.2.1 Preparing fields for fertilizer application

Fields should be prepared for fertilizing as it improves the growth of crop. For preparing the land follow the steps:

- Initially do the soil testing of the land to determine soil's level of nitrogen, phosphorus, and potassium and soil's pH.
- Remove large stones to make the field easier for tillage.
- Loose the soil by pitch fork or shovel by digging 12–14 inches in ground and then turning the soil over for loosening.
- Use recommended organic material over the surface of the soil and with tiller or pitchfork, evenly distribute the area.
- Using tiller or pitchfork, mix the compost with the soil so that the soil can benefit from the added nutrients in the compost.
- Use tiller lightly across the surface of the soil and organic material to level out the soil which makes easier for planting.
- Now apply fertilizer for planting of crop.

4.2.2 Mixing fertilizers

- A little fertilizer at or near planting time can help jumpstart seed toward a successful crop.
- When applying a fertilizer for seeds, application methods and rates are much more flexible with phosphorus than nitrogen.
- Wheat plants normally show significant increase in falling of tillers and better root development with certain fertilizers like phosphorus and nitrogen.
- When applying fertilizer with seed, nitrogen and potassium rates should be limited to avoid potential toxicity to the seedling.
- When placing fertilizer in direct contact with seed, cultivators should apply following guidelines:
- In 15-inch spaced rows, use fertilizer not more than 14 kg of nitrogen with potash for medium to fine textured soils or 9 kg for sandy or dry soils.
- In 10-inch rows, use fertilizer not more than 22 kg of nitrogen with potash for medium to fine soils and 12 kg for sandy or dry soils.
- For 6- to 8-inch rows, use fertilizer not more than 25 kg of nitrogen with potash for medium to fine soils and 18 kg for sandy or dry soils.

4.2.3 Apply fertilizers in right dosage, time and intervals of crop growth

- Timing of fertilizer application has a significant effect on crop yields.
- Proper timing of the fertilizer application increases yields, reduces nutrient losses, increases nutrient use efficiency and prevents damage to the environment.

- Applying fertilizers at wrong time might result in nutrient losses, waste of fertilizer and even damage to the crop.
- Plants need different nutrient rates and ratios at different growth stages.
- To make the nutrients available when plant needs them, fertilizers should be applied at the right timing.
- The optimum timing for fertilizer application is determined by Nutrient Uptake Pattern of the crop.
- For similar crop, each nutrient has an individual uptake pattern.
- The maximum rate of fertilizer that can be applied at one application depends on the salinity threshold that the crop can tolerate.
- Split fertilizer applications helps in avoiding salt damages to the crop and improving germination rate.
- Soil type affects the timing and frequency of fertilizer application where Cation Exchange Capacity and Soil Texture properties determine the frequency and timing of application.
- If nitrogen is applied too early before need of plant, then a required portion may get lost before the crop takes it up, so time nitrogen is in soil before the plant takes up should be minimized.

4.3 Study various diseases and insect pests and their infestation in seed production

4.3.1 Insects/Pests

Some insects are associated with damage to seeds or small seedlings. Not all pests are equally injurious and therefore the decision to control should be made individually for each pest species.

Greater False Wireworm

- The greater false wireworm occurs sporadically in summer rainfall region.
- It is the larval stage of dark-colored beetles with long legs, used for running on soil surface.
- In this, larvae can grow to ca. 35 mm in length and are smooth, hard-bodied with polished cream to golden-brown appearance.
- In this, larvae feed on seed and stems of seedlings below soil surface, making stem to have tattered appearance.
- The adult beetles damage emerging seedlings.
- In this, cultural practices supporting germination and rapid seedling development shorten vulnerable damage period of plant which limits seedling loss which retains plant densities.

Lesser False Wireworm

- The lesser false wireworm occurs sporadically all through summer rainfall region.
- It is larval stage of dark-coloured beetle known as dusty surface beetle.
- In this, larval stage is thin, with 10-15 mm long having dark shiny appearance.
- The larvae are very active and moves away in case of any disturbance.
- Also, the larvae feed on seed and stems of seedlings below soil surface, giving stem as tattered appearance.
- Cultural practices that supports germination and rapid seedling development shortens damage period of plant making limiting seedling loss and retaining plant densities.
- Targeting of larval stage in soil by seed treatments can be used where seedlings grow actively under moist soil conditions.

Black Maize Beetle

- The black maize beetle occurs all through summer rainfall region.
- They have shiny black colour and are 12-15 mm in length.
- They are found in or on soil surface layer.
- In this, larvae are grub-like having 25 mm in length, soft-bodied and white with darker head.
- The adult beetles feed on seedling stems below soil surface.
- Big holes with ragged edges can be seen after feeding, resulting in plant death and seedling density-loss.
- In this, larvae feeds on decaying organic material and will not damage the crop.
- The cultural practices that supports germination and rapid seedling development shorten damage period of plant and limits seedling loss by retaining plant densities.
- In this, chemical seed treatments are registered as pre-plant approach toward control of adult beetles.

Western Province Grain Worm

- It is a sporadic and unpredictable pest that occurs under intense rainfall conditions.
- It has C-shaped larvae and are legless and be 10 mm in length.
- It is white to cream in colour with small shiny brown heads.
- The brown-black adult beetles are called as weevils which is around 5-6 mm in length.
- The bodies of beetles are faintly spotted on back which shows characteristic snout on front of head.
- In this, soft white pupae can be found in cells in the soil.
- The larvae feeds on stems of seedlings just below soil surface, causing stem to have tattered appearance.

- The cultural practices supports germination and speedy development of seedling that shortens damage period of plant which limits seedling loss and retaining plant densities.

Sandveld Grain Worm

- The worm occurs in Western Cape.
- In this, larva is cream-coloured with dark brown head which is 10-12 mm long.
- The adult beetles are blue-green to metallic green of 6 mm long.
- In this, pupae are present inside cells below the soil surface.
- The larvae feeds on stem of plant below the soil surface causing small plants to die.
- Big plants survive feeding damage but tillering of these plants is reduced.
- The cultural practices supports germination and rapid seedling development to shorten damage period of plant and restricts seedling loss, thereby retaining plant densities.

4.3.2 Seed Diseases

Types

Fungal Diseases

- Fungi are small organisms made of filaments called hyphae.
- Most fungi use nonliving plant or animal material for nutrients and serve an important role in recycling organic matter in the environment.
- There exists small group of fungi which get their nutrition from living plants, resulting in plant disease such as pathogens.
- Pathogens are common and destructive of all infectious plant diseases.
- Many fungi reproduce by microscopic spores that are allowed to spread by wind, rain, insects, seeds, farm equipment, and runoff water or soil.
- Many plant-pathogenic fungi cannot survive in soil for long periods and can be effectively managed by clean plowing and crop rotation.
- Fungi such as most root and stem rot pathogens are soil inhabitants and survive well despite of efforts to starve out by crop rotation.
- Seed treatments can be used to control or suppress many seedborne, soilborne, and early-season foliar fungal diseases.

Bacterial Diseases

- Bacteria are microscopic, single celled organisms that reproduce by division.
- Like fungi, many bacteria are beneficial, with only few types causing plant disease.
- Bacteria cannot move on their own and depend on wind, animals, insects, farm equipment, seed, splashing rain, and other means to get to host plants.

- ❖ It can only enter plant by wounds or small natural openings, where they cause local or systemic infections.
- ❖ The symptoms of bacterial infection include wilting, soft rots, leaf blights, and spots.
- ❖ Many different bacterial pathogens can be seed borne while some bacterial diseases can be controlled or suppressed, directly or indirectly with seed treatments.

Viral Diseases

- ❖ Viruses are submicroscopic infectious particles that can be seen only with electron microscope or identified using specialized testing methods.
- ❖ They cannot move on their own and are spread by insects like aphids and leafhoppers, by seed and plant propagation.
- ❖ Viral pathogens such as wheat soilborne mosaic virus and wheat yellow mosaic virus are vectored by soilborne fungi.
- ❖ Viral diseases are systemic having symptoms such as discoloration, stunted or unusual growth, and poor development.

Examples:

Powdery mildew

- ❖ It is a white, powdery patches which is formed on upper surface of leaves and stem.
- ❖ It is a form of grayish white powdery growth that is present on the leaf, sheath, stem and floral parts.
- ❖ The powdery growth later on turns to black lesion and causes drying of leaves and other parts.
- ❖ It is a fungus which remains in high hills during summers in infected plant debris as dormant mycelium and asci.
- ❖ It is mainly spread by asciospores and through airborne conidia.
- ❖ It infects the plants during the period of high humidity and cool to moderate temperatures.

Loose smut

- ❖ It is a seed borne disease which results in infection during flowering by wind-borne spores.
- ❖ In this, the infection remains dormant inside the seed but the plants growth bears with infected inflorescence.
- ❖ In this, entire inflorescence commonly affects and appears as a mass of olive-black spores, which is covered by thin gray membrane.
- ❖ On rupturing of membrane, the head appears powdery.
- ❖ It causes infection by sowing infected seeds.
- ❖ Infection is favored by cool, humid conditions during flowering period of host plant.

Brown rust

- ❖ It is present on the upper leaf blades and also on sheaths, glumes and awns.
- ❖ In this, pustules are circular or elliptical and are smaller than stem rust which are not coalesce, but contain masses of orange to orange-brown Urediospores.
- ❖ The pathogen is found in summers in low and mid altitudes of Himalayas and Nilgiris.
- ❖ In this, the infection is started from wind deposits urediospores in eastern Indo-gangetic plains in mid of January where it multiplies and moves westwards by March

Stripe rust /Yellow rust

- It mainly takes place on the leaves and then on leaf sheaths and stem.
- It shows presence of bright yellow pustules which appear on leaves at early stage of crop arranged in linear rows as stripes.
- In this, stripes are of yellow to orange yellow in colour with teliospores which are dull black in colour.
- The pustules of stripe rust with yellow to orange-yellow urediospores forms narrow stripes on the leaves.
- The pustules is also found on leaf sheaths, necks and glumes.
- In this, inoculum survives in form of uredospores /teliospores in northern hills at off season on self sown crop or volunteer hosts that shows good source of inoculums.
- The disease is high when temperature is 10-20°C and very humidity.

Black rust

- It is found on almost all aerial parts of wheat plant and are common on stem, leaf sheaths, upper and lower leaf surfaces.
- The pustules are dark reddish brown which occurs on both sides of leaf, stems, and on spikes.
- Before spore masses break by epidermis, the infection sites feel rough and after breaking of spore masses, the surface tissues take on ragged and torn appearance.
- It mainly spread through uredospores from southern hills
- The disease is common in moisture and temperature above 20°C.

Flag smut

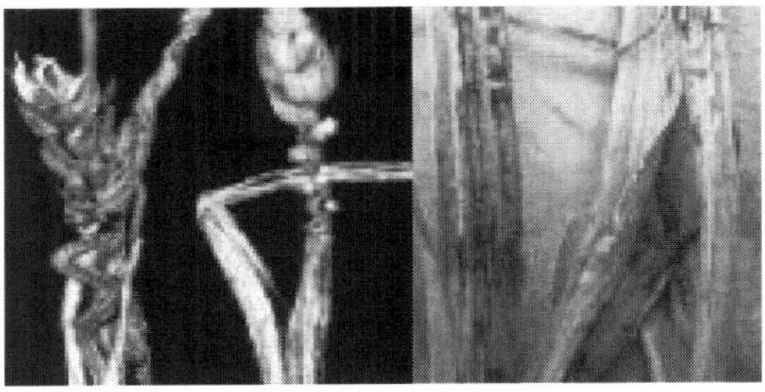

- ❖ Its symptoms is seen on stem, clum and leaves from late seedling stage to maturity.
- ❖ The seedling infection lead to twisting and drooping of leaves followed by withering.
- ❖ Grey to grayish black sori takes place on leaf blade and sheath which carries black powdery mass of spores.
- ❖ It is a seed and soil borne disease.
- ❖ The smut spores are viable for more than 10 years.
- ❖ The main infection occurs by sowing infected seeds or by resting spores present in the soil.
- ❖ The disease is common in temperature of around 18-24°C and also when relative humidity is 65% or above.

Hill bunt/Stinking smut

- ❖ It is a fungus which attacks the seedling of 8-10 days old and became systemic and grows along the tip of shoot.
- ❖ During flowering hyphae concentration in inflorescence and spikelets it transforms ovary in smut sorus of dark green color with masses of chlamydospores.
- ❖ When the diseased plants mature early, it will affect all the spikelets.
- ❖ In this, pathogen survives in seeds and sows the seeds which serves as a source of primary infection.
- ❖ The disease is common in temperature of around 18-20°C and also in high soil moisture.

Karnal bunt

- ❖ It symptoms are difficult to distinguish in field as incidence of infected kernels on a given head is low.
- ❖ It results as spreading of glumes due to sorus production and is not extensive as that of common bunt.
- ❖ The symptoms of it are most readily detected on seed after harvesting.
- ❖ It is a seed borne disease and sowing of infected seeds serves as main source of infection.
- ❖ The disease is common in temperature of 18-20°C and also in high soil moisture.

Leaf blight

- It is a reddish brown oval spots that appear on young seedlings with bright yellow margin.
- In severe cases, many spots coalesce which causes leaves to dry.
- It is a complex disease which mainly spread by externally seed-borne and soil borne conidia.
- The disease is common in temperature of 25°C and high relative humidity.

Foot rot

- The disease mainly occurs in seedlings and roots which turns rootlets to brown.
- It lead to pale green seedling with stunted growth.
- It is a fungus which produces sporangia and zoospores and oospores.
- The disease is soil borne, where pathogens survives in soil.
- Its mainly spread through soil and irrigation water.
- The disease is common in wet weather and high rainfall areas.

Head scab/Fusarium leaf blotch

Leaf blotch:

- The blotching caused by the organism is evident on leaves at about late joint to early boot growth stage.
- The young lesions occur as oval to elliptical, grayish green mottled areas, which is located where the leaf bends.
- The lesions enlarge rapidly and develops into large eyespot blotches with bleached or light grey centers.
- In this, the leaves tend to split or shred, beginning at the centers of the lesions.

Head Scab:

- It is caused by fungus.
- Its symptoms include tan or light brown lesions with one or more spikelets.
- Some diseased spikelets have dark brown discoloration at the base and an orange fungal mass along lower portion of glume.
- In this, grain from plants by Fusarium head blight is shriveled with white chalky appearance.

- Infected florets becomes slightly darkened and oily in appearance.
- The disease is soil borne and inoculums of fungi survive in soil.
- Spores are produced on crop debris left on or near the soil surface.
- The spores are transmitted to leaves by the wind or by splashing rain.
- The disease is common in cool, moist weather.

Helminthosporium leaf blotch

- Lesions caused by the disease is oval in shape and are dark brown in color.
- On maturing of lesions, centers turn light brown to tan color surrounded by irregular dark brown ring.
- It mainly affects on the lower leaves, starting as chlorotic flecks or spots.
- The infection sites enlarge, turn dark brown and coalesce.
- When the disease is severe, it affects the leaves or leaf sheaths and makes them to die prematurely.
- It is a seed and soil borne diseases in which the inoculums is present in seeds and soil as main source of infection.
- The disease is common in very humid and higher rainfall areas.

Seedling blight

- The symptom of the disease is seen when ears become infected during early flowering stages.
- Later infections result in infection of grain but without obvious bleaching of ears.
- Fusarium lesions started in leaf sheath at stem base where crown roots split the leaf sheath when emerging.
- The infection spreads in the leaf sheath which causes long dark brown streaks at stem base.
- The ear blight phase of disease causes yield loss and result in mycotoxin production in grains.
- An important source of Fusarium for wheat crops is seed, but fungus can also survive on debris in the soil.
- The disease is common in relatively high soil moisture and in high soil temperature.

4.4 Understand seed treatments and judicious use of insecticides/ fungicides and bio fertilizers.

4.4.1 Seed treatment

- It describes products and processes.
- The usages of specific products and specific techniques improve the growth environment for seed, seedlings and young plants.
- It shows complexity ranges from basic dressing to coating and pelleting.
- Due to infection in seeds, crop gets different diseases like: smut, rust, blight etc.
- It is used on many crops to control variety of pests and ensure uniform stand establishment by protecting against soil borne pathogens and insects.
- Seed treatments can be used to suppress root rots in certain crops.
- It can protect seed and seedling from attack by certain insects and pathogens.
- With such diseases, there is adverse effect on annual production of crop.
- The infection from diseases could be minimized or reduced by using only healthy, and quality seeds chosen through hot water seed treatment and treatment with fungicide.

Methods of seed treatment

(A)Hot water seed treatment

- Initially, soak the seeds in ordinary water for about 4 hours, than dip in hot water 52 °C for 10 minutes or 55 °C for 5 minutes.
- After that, cool and dry the seeds in shade, and let for fungicide treatment.

(B) Fungicides treatment of seed

- It is compulsory to treat the seeds for producing quality and healthy crop.
- Mix the wheat seeds before sowing with Bevistin or Benlate 2 grams (0.2%) per kg seeds in an air tight pot and stir it vigorously for 8-10 minutes.
- Treating the seeds this way helps in controlling diseases and germs on outer surface of wheat grain such as blast, loose smut, etc.

(c) Seed Dressing

- The most common method of seed treatment.
- The seed is either dressed with a dry formulation or wet treated with a slurry or liquid formulation.
- Dressings are applied both on-farm or in specialised seed treatment facilities.

(D) Seed Coating

- A special binder is used with a formulation to enhance adherence to the seed and begin to impact seed size and shape.
- Coatings require advanced treatment application technology.

(E) Seed Pelleting

- ❖ The most sophisticated seed treatment technology, resulting in changing the physical shape of a seed to enhance plantability and handling.
- ❖ Pelleting requires specialised application machinery and techniques and is the most expensive of the applications.

4.4.2 Biofertilizers

- ❖ They are carrier based preparations with effective strains of microorganism like Bacteria, Fungi, Algae in combination with numbers when incorporated with seeds, seedlings and in soils.
- ❖ They provide nutrients to the plant through microbial activity.

Benefits:

- ❖ Increase crop yield by 20-30%.
- ❖ Replace chemical fertilizers by 25%.
- ❖ Restore natural fertility.
- ❖ Provide plant nutrients at very low cost.
- ❖ Have no harmful effects on soil fertility and plant growth.
- ❖ Hasten seed germination, flowering and maturity in crop.
- ❖ Helps in recycling/ decomposition of organic waste.
- ❖ Provide residual effects for subsequent crops.
- ❖ Pollution free and eco-friendly.

PRECAUTIONS:

- ❖ Store Biofertilizers in cool and dry place.
- ❖ Open the packets just before use and all its content be used at time.
- ❖ Biofertilizers treated seed should not be mixed with chemical fertilizers, pesticides and insecticides.
- ❖ If seeds are to be treated with fungicides, treat the seeds first, then with double the recommended dose of Biofertilizers.
- ❖ Seed treatment and mixing of Biofertilizers with compost should be done in the shade.
- ❖ Used of Biofertilizer is being emphasized along with chemical fertilizers and organic manures. Biofertilizers are not replacement of chemical fertilizers but can supplement their requirements.

4.4.3 Fungicides

Treating seeds with fungicides is beneficial under following conditions:

- ❖ Early planting in cold, wet soils
- ❖ Reduced till and no-till fields

- Seed with low germination rate or low seed vigor

4.5 Understand various seed borne diseases and their management

4.5.1 Seedborne diseases

- Disease causing organisms may be carried in seeds in suitable environmental conditions, gets transmitted to cause disease in developing seedlings.
- Due to certain diseases, pathogen attacks germinating seedling and affects seedling establishment with others.
- In this, disease symptoms are not seen till later stage of growth.
- Due to seedborne infection, optimum control is achieved cost-effectively with least disruption, using seed treatment chemicals or by avoiding infection by producing seed in areas that are unfavourable for disease development.
- As observed, majority of seedborne pathogens are carried passively on seed surfaces and are established as hyphae or spores under seed coat.
- For certain plant diseases, seedborne inoculum is the source of infection, so if healthy seed is sown, disease will not develop.

Importance of Seed-borne Diseases

- Many microorganisms in natural environment exist in multi-cellular aggregates described as biofilm association with solid surface like seeds.
- The surface properties of plant tissues, nutrient and water availability and species of colonizing pathogens strongly influence the resulting biofilm structure.
- Seeds are regarded as highly effective means for transporting plant pathogens over long distance.
- The consumer demand for new plant products gave opportunity for many plant pathogens to disseminate to new areas on imported seeds.
- Many seed-borne pathogens induces rots on produce which reduces market value of affected produce, hamper addition of value and prevent produce to complete their roles in food chain.
- Controls are used which increase cost of production where costs are passed over to end users.
- The impact of seed-borne pathogens results in production of many types of mycotoxins on produce like oil seeds, maize and cereals.
- Mycotoxins are toxic, low molecular weight, secondary metabolites produce some species of fungi.
- Mycotoxins of agricultural importance are aflatoxins, fumonisins, ochratoxin, zearalenone and deoxynivanelol.
- Fungal toxin contamination of food products causes acute or chronic intoxications, leading to reduced life expectancy; exacerbate disease conditions which shows 40% loss of economic productivity.

4.5.2 Procedure/precautions for seed treatment for control of borne diseases

The following things need to be considered while treating the seeds:

- ❖ Select non-windy place to treat seeds with fungicide so that air will not carry away the fungicide.
- ❖ Children, aged people, pregnant women are more vulnerable to poisonous effect of fungicide hence; they should not be involved in treating the seeds.
- ❖ Use mask, gloves, and goggles while treating seeds.
- ❖ Wear the clothes that cover whole body.
- ❖ Use only proper or prescribed amount of fungicides.
- ❖ Wash thoroughly with water when fungicide touches your skin directly.

Protection of Seeds and Seedlings

- ❖ Seeds and seedlings are vulnerable to many soil borne and foliar pests.
- ❖ Insects and pathogens can destroy germinating seeds and young plants, which are relatively tender and lack food reserves to recover from injuries or to survive extended periods of stress.
- ❖ Under stressful conditions, a number of aggressive or even fairly weak pathogens can become active and cause plant population and yield losses.
- ❖ Seed treatments can protect the seed and seedling from attack by certain insects and pathogens. No systemic fungicides or insecticides form a chemical barrier over the surface of the germinating seed which protect the germinating seed from chewing insects, such as wireworms, or soil borne pathogens, such as pythium.
- ❖ Certain systemic seed treatments can protect aboveground parts from sucking insects, such as aphids, or foliage diseases, such as rust.
- ❖ Systemic fungicides and biological seed treatments can also protect young plants from root rot. Although the duration of protection may be limited, a delay in infection can reduce the losses. For chronic diseases, such as root rots, the earlier that the infection takes place, the greater will be the damage.
- ❖ Seed treatments lasts only for 10 to 14 days beyond planting, with pesticide breakdown being most rapid under warm, moist conditions.
- ❖ Certain active ingredients can protect seedlings considerably longer when applied at the highest labelled rate

Advantages

- ❖ Seed borne pathogens are vulnerable: The seed borne phase is weakest link in life cycle for many plant pathogens. With seed treatments to control seed borne pathogens is very effective for disease control.
- ❖ Precision targeting: Seed treatments are not subject to spray drift as chemicals are applied directly to seeds, little is wasted on no target sites like bare soil.
- ❖ Optimum timing: Seeds and seedlings are vulnerable to diseases and insects as

compared to mature plants. Applying treatments to seeds will allow pesticides to be present when needed most.

- Low dose: Relatively small amounts of pesticides are used in seed treatments compared to broadcast sprays as it reduces the cost and the potential environmental impact and the probability of chemical residues in harvested grain.
- Easy to apply: Seed treatments are relatively easy and cheap to apply as compared to broadcast sprays.

Disadvantages

- Accidental poisoning: Treated seed looks like food to some animals as hungry livestock birds, such as pheasants or quail consume spilled treated seed.
- Cropping restrictions: Just like other pesticides, some seed treatments may have significant grazing or rotation crop restrictions.
- Limited dose capacity: The amount of pesticide that can be applied is limited by how much will actually stick to the seed. Seed coating technologies are helping to overcome this limitation, but phytotoxicity may still be a problem.
- Limited duration of protection: The duration of protection is short due to relative small amount of chemical applied to seed, dilution of chemical as plant grows, and breakdown of chemical.
- Limited shelf life of treated seed: Producing excess treated seed is undesirable as shelf life of treated seed is limited. Surplus treated seed cannot be sold for grain which is a serious limitation for seeds where seed germination and vigor decline relatively quickly.
- Phytotoxicity: Pesticide injury to plant tissues is called phytotoxicity as seed treatments exist in high concentrations on tender tissues of germinating seeds and seedlings having very low phytotoxicity.
- Worker exposure: In course of treating and handling large volumes of seed, workers may be exposed to seed treatment chemicals as aerosols. Inhalation of aerosols and skin contact with seed treatments must be prevented in the seed treatment process

Unit 5

Seed Harvesting and Threshing Management

5.1 Understanding proper stages of crop maturity during seed production

5.1.1 Introduction

- Harvesting at correct time is crucial for quality development after harvest and for attaining longest postharvest life.
- Maturity is a natural phenomenon which is considered as stage of development, where plant/crop superimposes and shifts from vegetative to reproductive stage.
- Harvesting of horticultural produce at right stage of maturity is important for optimum quality in order to maintain further intact intrinsic quality for maximum returns.
- Maturity is a stage of full development of tissues of fruit only after which it will ripen.
- Maturity stage at harvest will not have an effect on fig weight, size, antioxidant capacity, skin color and thickness, split ostiole and blemishes
- In case of fruit crops, adequate maturity and harvesting indices are features or components that change during development, correlating adequately and consistently with quality, without affecting by environmental factors.
- Maturity in crops can be divided as physiological maturity that shows period when sexually induced reproductive growth ceases while harvest maturity shows where seed, fruit or other economically important organ of yield reaches at state of ripeness that can be removed from parent plant for consumption.

5.1.2 Estimation of Maturity

- To up keep the quality of product
- To enhance freshness/appearance /elegance of the produce
- Improvement in the storage life of the produce
- Management of ripening and senescence (hasten /delay)
- Extended utilization of the produce

- Easy of handling
- To maximize returns
- To manage the environmental factors
- To manage pest and diseases
- For long distance transportation of the produce

5.1.3 Maturity Indices

- Maturity as stage of development gives minimum acceptability quality to consumer showing measurable point in commodity's development, which shows need for techniques to measure maturity.
- Maturity index for commodity is measurement that can be used to show whether particular example of commodity is mature. Such indices are important to trade in fresh fruits and vegetables for many reasons.
- Maturity measures by producers, handlers and quality control personnel should be simple, readily performed in field which requires relative inexpensive equipment.
- The index is preferably an objective rather than subjective,
- The index must consistently relate to quality and post harvest life of commodity for all growers, districts, and years.

5.1.4 Strategy for maturity index

The strategy for developing a maturity index is to:

- show changes in commodity all through its development.
- look for features such as size, color, solidity, etc. whose changes correlate well with stages of commodity's development.
- use storage trials and organoleptic assays to show the value of maturity index which shows minimum acceptable maturity.
- assign index value for minimal acceptable maturity when relationship among changes in maturity index quantity and quality and storage life of commodity has been determined.
- test index over many years and in many growing locations to ensure of consistent reflection in quality of harvested product.

5.1.5 Stages of Maturity Crop

The type of crop, climatic conditions and humidity of material has influence on correct election of moment of cut. At that time, it is necessary to keep fluid dialog with producer which needs follow up of crop and notification of maturity state as per given climatic conditions. In terms of humidity of crop:

- Milky Grain: very green plant, low dry matter, likely fermentation problems draining of nutritious effluents, easy compaction.

- Doughy Grain / Full: it is the moment of highest forage volume with the best quality in the composition of the silo due to the excellent relation ear/plant. There may be one or two inferior leaves completely dry.
- • Hard Grain: it has very little green volume and low quality due to the high deterioration of the leaves and stems; it causes problems with the chopping, and the compaction; the animal does not cash in on the grain because it is too hard. In this situation, it is important to raise the height of cut and recommend the use of Corn Cracker to crack the grain very well and take advantage of all its nutrients.

5.2 Role of seed moisture during harvesting and threshing of seed crops

5.2.1 Seed moisture

- The moisture content is amount of water in the seed which is expressed as percentage.
- The moisture inside seed can be expressed on net weight basis or dry weight basis.
- Seed moisture content can be accurately determined experimentally by scientific techniques, or can be predicted from information.
- The determination is destructive to seeds used, as in many cases, prediction is sufficient, so determination should only be done where required.
- The moisture content inside the seed has large effect on storage life of seeds, hence it is important to know the moisture content in order to make an accurate prediction about storage life of every accession.

Predicting seed moisture content

Measure the environmental conditions

- Measure relative humidity of atmosphere where the seeds have been held for past few days using aspirated hygrometer.
- Measure mean temperature where the seeds have been held for past few days.
- If seeds are from different environment, they need to be allowed to equilibrate with current one before making any predictions.

Checking equilibrium moisture content

- Obtain the name of crop in left hand side of tables of equilibrium moisture contents.
- Read equilibrium seed moisture content at relative humidity measured as shown above.
- Use the figure of approximate moisture content, since predicted value in making estimates of drying period or exact value of moisture content is not necessary.

5.2.2 Seed moisture during harvesting of seed crops

Seed moisture during harvesting

- Seed moisture content is most reliable indicator of seed maturity and harvest timing in seed crops.

- There are significant times during harvest as knowledge of seed moisture is critical: at swathing and combining.
- Swathing in correct range of seed moisture content maximizes seed yield and minimize seed losses during the harvest.
- Taking accurate measurement of seed moisture content is main component of economic seed crop management.

5.2.3 Seed moisture during threshing of seed crops

Threshing

- Threshing detach the grain from panicles which can be done by rubbing, stripping, impact action or using all actions.
- Threshing operation can be done manually (trampling, beating) using animal power or mechanical threshers.
- Manual threshing is common practice in developing countries.
- Grain spillage, incomplete separation of grain from chaff, grain breakage due to excessive striking, are main causes for loss during threshing process.
- Thresher, which is a machine that separates grain from harvested crop provides clean grain without much loss and damage.
- In threshing, seed loss due to broken seed, unthreshed seed, blown seed, spilled seed etc. can be minimum.
- Clean unbruised seed fetches good price in the market and has long storage life.
- Threshing by bullock treading is practiced on large scale in the country but it is also time consuming and involves drudgery. Tractor in many places is now used in place of animals for treading.
- Methods of threshing influence seed quality in terms of germination and vigour.
- The different threshing methods produces breaks, cracks, bruises and abrasions in seeds which in turn results in abnormal seedlings of questionable planting value.

Impact of threshing and moisture content on seed quality

- The required threshing moisture for certain seeds should be around 14%-16%, so threshing above moisture content will show deterioration in seed quality.
- The mechanical damage to seed during threshing varies as per genetics of variety.
- Seed hardness plays an important role in seed processing and vitreous endosperm variety pose breakage resistance.
- The percentage of damaged seed using wooden stick was higher and different from threshed using paddy thresher.
- The percentage of seed germination was higher using paddy thresher and different from threshed using wooden stick.

5.3 Causes of mechanical mixture during harvesting and threshing

5.3.1 Mechanical Mixtures

- These are an important reason for varietal deterioration which occurs at the time of sowing if more than one variety is sown with same seed drill, by volunteer plants of same crop in seed field, or by different varieties grown in adjacent fields.
- Two varieties growing next to each other field is mixed during harvesting and threshing operations.
- The threshing equipment is contaminated with seeds of other varieties.
- The gunny bags, seed bins and elevators are contaminated by adding to mechanical mixtures of varieties.
- Roguing the seed fields critically and using utmost care during seed production and processing are required to avoid mechanical contamination.

5.3.2 Avoiding Mechanical mixtures

- It is an important source of variety deterioration at the time of seed production.
- It results at the time of sowing when more than one variety is sown with same seed drill using volunteer plants of similar crop in seed field or by different varieties in adjacent fields.
- The seed produce of all the varieties are kept on same threshing floor, resulting in considerable varietal mixture.
- Avoiding of mechanical contamination can be done by rogue the seed fields and practice the utmost care at the seed production, harvesting, threshing and further handling

5.4 Various methods of harvesting and threshing and handling of harvested and threshed produce.

5.4.1 Introduction

Harvesting of crop is difficult process as it requires good preparation and timing. If dry crop is left out in fields for longer time, winds and storms can destroy the crop. The quality of the seed crop may decrease if seeds gets rained on which dries again. Harvesting of seed requires use of combine- a heavy machine.

Harvesting of seed cops started as soon as crop field gets dried enough so that it can be handled safely. For checking, use moisture meter to know quick determination about crop condition. If meter is not available, then weigh 0.25- to 0.5-kg of sample, dry it at 260°F and reweight it again.

5.4.2 Methods

There are different methods of harvesting of seeds which are followed in different countries, as it mainly depends on timely labour availability and costs associated with harvesting. Harvesting of seed crop can be done by:

- Manual

- ❖ Animal drawn machine
- ❖ Mechanically operated machine.

Manual Harvesting

Till 19th century, grain were harvested by cutting with sickle, scythe, cradle or manual mowers.

- ❖ It is the main method of harvesting in developing and under-developed countries.
- ❖ In developed countries, certain crops are harvested manually for high quality and premium price.
- ❖ In this, the crops are handpicked without using any tools.
- ❖ The sickle, spade and digging rods are main traditional tools for harvesting and digging while other tool which is used for harvesting is knife where output is low and effort required is high.
- ❖ In inderminate crop, harvesting is done at intervals as economic product maturity results at different periods.

Sickle

Use of Sickle

- ❖ It is one of the common hand tools used for harvesting of crops.
- ❖ It consists of curved steel edged blade fixed to wooden handle.

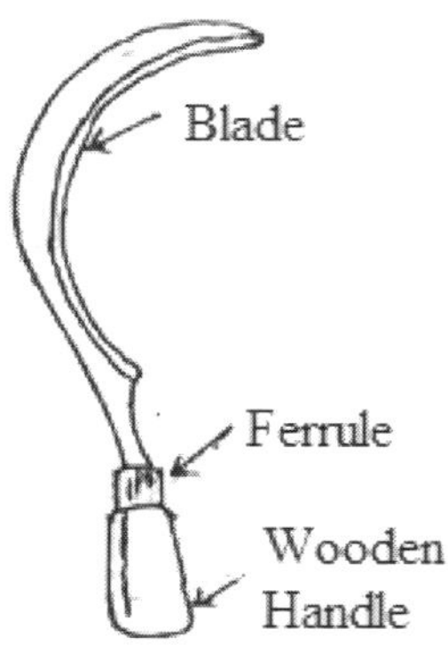

- ❖ The cutting edge can be plain or serrated, depending on its design.
- ❖ The curvature of sickle blade differs in plain edged and serrated sickles.

- The sickle blade is fabricated from medium carbon or high carbon steel.
- The blade has a knife section where its thickness reduces towards the cutting edge.
- The tang of the blade is inserted in wooden handle for joining the blade to the handle.
- The wooden handle is straight or bent at the end.
- In this, cutting is done by translatory and rotary movement to the blade around the point of cut.

Scythe

- It is a manually operated having a shaft of about 170 cm long known as snaith made of wood or metal.
- Snaiths are straight with offset handles having one or two short handles at right angles to it, usually one near the upper end and always another roughly in the middle.
- In this, handles are adjustable to suit the user.
- A curved, steel blade between 60 to 90 cm long is mounted at lower end at 90° or less to snaith.
- In this, the blade is projected from left side of snaith when in use, with the edge towards the mower.
- The mower holds the top handle in left hand and central one in right, with the arms straight with blade parallel and close to the ground and uncut wheat to the right.

Cradle

- It is a more developed tool for reaping and harvesting.
- It comprises of arrangement of fingers that are attached by light frame work which snatches of scythe.
- It gather straws as it is cut and deposit it in a swath.
- After cutting stroke, cradle is tilted to drop the straw in a pile.
- It can be useful in cutting more grains in a day.

❖ In this, due to more shattering loss, sickle is used to harvest the grains.

Reaper

❖ Reaper is a harvesting-machine used for cutting grain-crops.

❖ It delivers grain to one side in gavels which can be ready to bind into sheaves and/or elevate the gavels on platform where two operators bind them into sheaves by hand.

❖ Mechanical reapers and binders were developed to cut and windrow grain for field drying.

Self propelled vertical conveyor reaper

❖ It carries crop row divider, star wheel, cutter bar and lugged canvas conveyor belts.

❖ It cut the wheat crops and conveys it vertically to one end and windrows the crops on ground.

❖ It is a vertical type conveyor reaper which is walking type, self-propelled riding type and tractor mounted type.

Mechanical harvesting machines

Harvesting of grains by machines is an important part of mechanized agriculture. Power operated machines can be reaper, mower, rotary power scythes, forage harvester, binder and conventional combines.

The machines used for efficient harvesting of seed crops include:

❖ Reaper binders

❖ Combine harvesters

❖ Tractor front and rear mounted reapers

1. Reaper binders

- ❖ It comprises of cutting unit as disc type or cutter bar type.
- ❖ After cutting, crop is conveyed vertically to binding mechanism which is released to ground in bundles.
- ❖ It is basically used for different kinds of low stem crops, rice and wheat.
- ❖ It can be used in different types of land forms like hills, slops, mountains, etc.
- ❖ In this, steer control handle bar is freely adjusted to up, down, left or right.
- ❖ It has a shaft drive system which works more stable and safe.
- ❖ It can be used on mountains or areas where combined harvesters cannot easily work.

2. Combine Harvester

- ❖ It is a machine which harvests grain crops such as wheat, oats, rye, barley, corn (maize), sorghum, soybeans, flax (linseed), sunflowers, peas, beans, lentils and canola.
- ❖ It is a combination of harvesting—reaping, threshing and winnowing.
- ❖ It carries hydrostatic drives.
- ❖ In this, cutter bar runs the full length of header below the reel where its teeth gets open and close repeatedly to cut the crops from its base.
- ❖ Here, behind cutter bar, cut crops get fed toward the center by spinning augurs which travels up conveyor to processing mechanism in main part of combine.
- ❖ In this, unwanted chaff and stalks, passes along conveyors toward the back of the machine.
- ❖ Although the greatest application of combines is in harvesting the small grains, corn and soybeans, these are also used for wide variety of small-acreage or specialty crops.

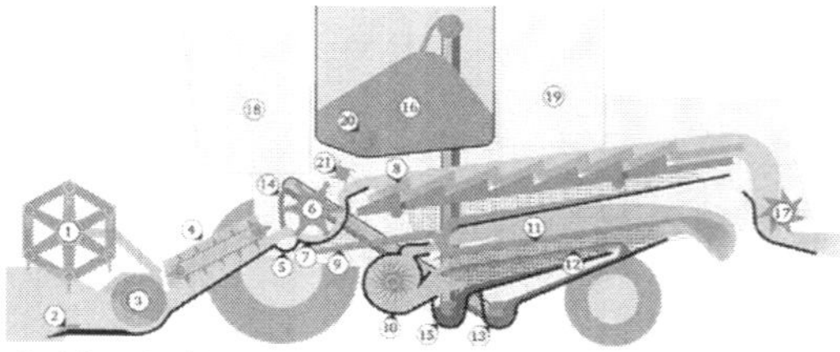

1) reel; 2) cutter bar; 3) header auger; 4) grain conveyor; 5) stone trap, 6) threshing drum; 7) concave; 8) straw walker; 9) grain pan; 10) fan; 11) top adjustable sieve; 12) bottom sieve; 13) tailings conveyor; 14) rethreshing of tailings; 15) grain auger; 16) grain tank; 17) straw chopper; 18) driver's cab; 19) engine; 20) unloading auger; 21) impeller

Its main function includes:

- ❖ Cutting the standing crops
- ❖ Feeding the cut crops to threshing unit
- ❖ Threshing the crops
- ❖ Cleaning the grains from straw
- ❖ Collecting the grains in a container

3. Small scale harvester

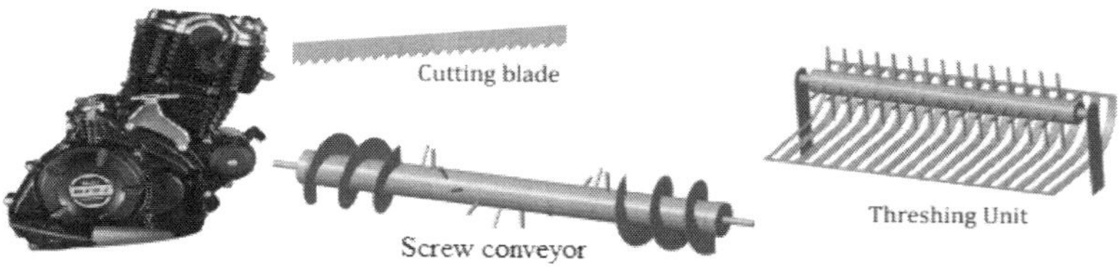

- ❖ It comprises of four stroke single cylinder petrol engine.
- ❖ It has a cutter blade assembly with sliding and stationery cutter blade where cutters are triangular in shape.
- ❖ In this, the screw conveyor collects the crops and transfers it for further operation.
- ❖ In this, the threshing unit threshers the dry crop which separates small grains from their chaff and straw.

4. Tractor mounted vertical conveyor reaper

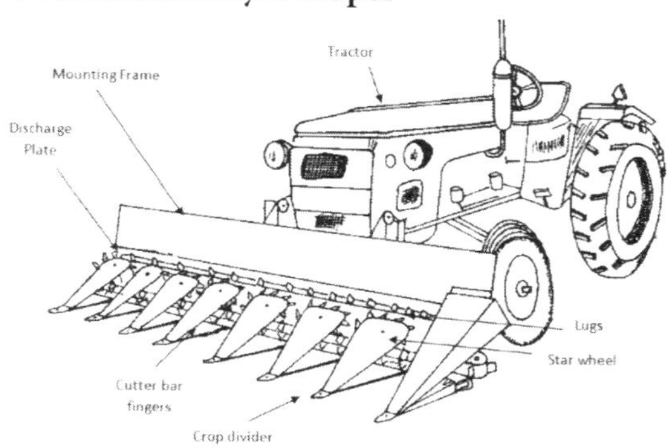

- It comprises of reciprocating cutter bar assembly, seven crop row dividers, two vertical conveyor belts fitted with lugs, pressures springs, pulleys and gearbox for power transmission system.
- In this, crop row dividers are fitted in front of cutter bar assembly.
- The star wheels are mounted over the crop row dividers.
- The machine is kept in front of tractor where the power is given from tractor PTO using intermediate shaft that runs below the chassis of tractor and coupling shaft.
- Here, the height of machine which is above the ground is controlled by tractor hydraulic using pulleys and steel ropes.
- Once the crop is cut by cutter bar, it is held in vertical position and delivered to one side of machine by lugged belt conveyors which falls on ground as windrow perpendicular to machine movement direction.

5.5 Impart knowledge on seed contaminants during harvesting and threshing

5.5.1 Seed contamination

- Seed contamination is a critical issue in controversy in terms of genetically modified versus non-genetically modified organisms.
- It is considered an important mechanism by which herbicide resistance can be transferred from farm to farm.
- When seeds of weed species not previously found in a region are used as contaminants during trade in plants and plant products, seed-borne viruses exists.
- It can introduce viruses that invade native vegetation, and infects newly introduced weeds by viruses emerging from indigenous vegetation.
- It provides attractive hosts to new arthropod vectors.

5.5.2 Rules

Seed contamination can be identified by inspection of consignment as per laboratory analysis of submitted seed sample from seed lot. The seed inspection needss varies as per seed lot size and size of seed.

For seed lots less than 10 kg OR seed lots with seed greater than 8mm diameter:

- Seed lot is subject to inspection by biosecurity officer on arrival.
- Seeds will be inspected visually to assess contamination.
- If contamination is found, sample must be drawn as per Seed Testing Agencies procedures.

For seed lots greater than 10 kg and containing seed less than 8mm diameter:

- Seed lot to be sampled as per Seed Testing Agencies procedures and methodology, and sample(s) forwarded to departmental laboratory for analysis.
- Results of seed analysis must be presented officially with certificate.
- Certificate will outline contamination presence in sample.

Unit 6

Post Harvest Management of Seeds

6.1 Post harvest management: farm storage/seed drying/conditioning/ pre-cleaning and cleaning/grading/ seed treatment/packing and bagging.

6.1.1 Farm storage

Introduction

- ❖ Nearly two-third of total produce is retained at farm level for food, feed and seed.
- ❖ Majority of farmers uses traditional type of storage which is open to damage and loss from various sources.
- ❖ Seed grain storability depends on:
    - moisture content
    - temperature
    - initial condition
    - insects and molds
- ❖ The main problems with existing facilities is short life of material used, not safe from weather effect, rodent and inadequate capacity.
- ❖ Majority of cultivators in all holding size classes, do not adopt improved storage facilities due to lack of funds, non-availability of desired type of container and in some cases unawareness about improved storage facilities.
- ❖ The farmers are aware of good quality storage structures and the precaution to be taken during storage as a of which, farmers adopted 'metal bin' for storing grains and oilseeds.
- ❖ Many farmers are following the method of open heaping and gunny bags for storage of seeds.

Method

For storage of seeds, the store must satisfy the following parameters:

- ❖ seed must be kept dry

- seed should be kept at uniform temperature
- seed should be protected from insect attack
- rodents and birds should be excluded

Storage Structure

- The mode of storage in dwelling houses is tin drums, big earthenware, pucca kutcha kothas, thekas, underground mud walled or concretized cellars and R.C.C. bins.
- Many indoor and outdoor structures are made of mud, wood, straw, jute bags, bricks and split bamboo.
- The traditional storage structures varies from region to region which are not rodent, moisture and insect proof and require frequent maintenance.
- Seed are traditionally stored in mud kothi, jute bags in rooms or buried under straw in underground structures.
- Safe storage period reported from 2 months to one year.
- There is a need to augment the production machinery to ensure appropriate and improved storage facilities like metallic bins, pusa bins etc are available to meet the needs of farmers in different holding size classes.

Disinfestations of store before storage of seeds

- Before use of godowns/stores for keeping seeds, it should be cleaned with dirt, rubbish, webbing and refuse materials.
- In godowns/stores, cracks, crevices, holes in wall, floor or ceiling should be filled with cement.
- Uneven floor, wall or ceiling in godowns/stores should be brought to level.
- The range of available insecticide products is broad enough to treat different surfaces as per their characteristics.
- It is essential to maintain the required concentration of fumigant in the sealed space and for sufficient time in order to have effective control of all stages of the pest.
- In case of uneven surfaces of godowns/stores, it should be treated by spraying with powder mixed with water.
- For smooth, non-porous surfaces of godowns/stores, spraying is done with stickier concentrate.
- Use of ambient treatment will help in destroying flying insects by aerosols in hermetically sealed premises which needs to be done mostly in evening when flying insects are active.
- Spraying do be done can be mechanical, pneumatic or thermal which provides better distribution of product over the grain.

Before storing of seeds, go-downs/store houses should be disinfected with methods of disinfestations as:

(a) Fumigation of empty store house with hydrogen cynide as it is widely used fumigant and is highly poisonous.

(b) DDT (dichloro/diphenyl/trichloroethane) and BHC (Benzene hexachloride) in form of smoke, dust or spray at concentration of 50-100 mg active ingredient per sq. meter.

(c) Malathion spray of around 100 mg/sq. m is better.

(d) Burning of sulphur.

(e) Good pyrethrum spray at about 6-8 oz/1000 c.ft is quite effective.

6.1.2 Seed drying

Drying

- Drying is a post-harvest system phase in which, the product is rapidly dried till it reaches safe-moisture level.
- The idea of drying the seed grains is to lower the moisture content in it for safe storage and further processing.
- Drying in growing countries is mostly done by spreading thin layer of seeds in sun, on the threshing floor or on rooftops.
- Mechanical drying of seed grain is not practiced in many developing countries.
- Seed are mostly sun dried, since sun drying depends on weather conditions leading to dirty grain, spillage loss and bird attack.

Methods of drying

Drying of seeds is done in two stages. The first drying is by delaying the harvesting of seed crop while second drying involves windrowed or bundled of harvested seed crop which is left in the field or transported to threshing site and stacked for further drying.

For drying seed grain, normally two methods are used:

- natural drying
- artificial drying

Natural Drying

- In this, the seed grains are exposed to air (in sun or shade).
- For obtaining desired moisture content, seed grains are spread in thin layers on drying-floor, where it is exposed to air.
- In this, the duration depends on moisture content as required for safe storage.
- To get uniform drying, seed grains must be stirred frequently, especially if it is in direct sunlight.
- For effective drying, relative humidity of ambient air should not be more than 70%.
- Here, grains should not be exposed at night and the method should not be used in humid regions or during the rainy Seasons.
- The insufficient or excessively slow drying can led to severe losses in product during storage from self generated heat of green seed grain.

- Seed grains should be sun-dried in the middle of the day only and should not be dried in bags on a concrete floor.
- The process requires very low cost and has flexible capacity of grains o be dried up.
- It is a slow drying process which normally depends on the weather conditions.
- In this process, high risk of contamination with possible loss and risk due to birds and rodents.

Artificial drying

In this, heated air (dryers) or unheated air (dehumidifiers) is blown through a grain mass.

Forced hot-air Drying

- This is the most wide spread practice in semi-humid and humid conditions where natural drying cannot be used.
- Artificially heated air is forced to flow through a mass of seed grain in bulk or in bags to absorb released moisture from grain mass
- It facilitates early harvesting and reduces shattering losses in the field
- It is more rapid drying in terms of capacity and is independent of weather conditions
- It requires heavy initial cost with high cost of drying
- In this, there is risk of seed damage due to high temperature
- It needs skilled personnel and regular monitoring with risk of breakdown or fuel shortage at critical times.

Dehumidified air-drying

- In this, unheated dehumidified air is circulated through seed grain mass till moisture content of grain is reduced to required level.
- It is commonly applied when small quantities of seed grains are dried to very low moisture content for being used as seed.
- It is simple and easy to manipulate which involves low risk of seed damage from hot air
- It is not good for large-scale production as it involves high cost of investment

Types of dryers in artificial drying

For artificial drying of grains, two types of dryer are used :

Static or discontinuous dryers or batch dryers

- Such dryers are less expensive and can handle medium quantities of grain.
- Floor dyers : Suitable for all seeds of all sizes.
- Bag dryers : Suitable for small seed grain lots to be used as seed where many varieties are handled. All types of grains in sacks.
- Box dryers: Suitable for large seed grain lot where slow drying is required.

Continuous dryers

- These are high-flow dryers that need a more complex infrastructure, complementary equipment and special planning.
- They are appropriate for big centers, silos or warehouses, where very large quantities of product are handled.
- Continuously flowing vertical dryers: Good for quick and normal drying and free flowing grains of single variety which can handle large volume of grain and are not good for small lots of different varieties.
- Continuously flowing belt dryers: Good for quick and normal drying seeds, small to very small lots and chaffy seeds.
- Rotary dryers: Good for quick and normal drying grains, chaffy grains with moderate moisture which are normally for large capacity drying and not for grains with high moisture.

6.1.3 Pre-cleaning and cleaning

Pre-Cleaning of seeds

- The seed which will be graded or treated needs to be properly cleaned and conditioned by removing seed appendages and other materials from the seeds.
- Pre-cleaning of seeds on the farm is done to remove large unwanted materials like straw, large stones, etc.
- To pre-clean the seeds, grower/farmer can use equipments such as scalpers, debearders, huller-scarifier, Buckhorn machine and maize sheller..

Cleaning of seeds

- ❖ When a grower/farmer have completed the pre cleaning proocess of seeds, he/she needs to clean the seeds again.
- ❖ Cleaning is an important operation which helps in upgrading physical quality of seeds.
- ❖ In basic cleaning, broken and rotten seeds are removed along with any impurities which have been missed during pre-cleaning process.
- ❖ For cleaning of seed, seed size and weight of seeds are considered in seed cleaner equipments at farm.
- ❖ The basic principles which are used in cleaning includes aspiration, scalping and grading.
- ❖ Grower/gardener can use air screen cleaner equipment for seed cleaning which functions similar to blower, where there are four levels, that are used to screen seeds of four different sizes and in separation.
- ❖ For good cleaning process of seeds, grower/farmer should check for moisture content of seed and should see for seed bins, elevators, seed cleaners and gunny bags.

Method of Cleaning Seeds

Cleaning of seeds can be done by:

Preparing seeds for basic seed cleaning

- ❖ Pre conditioning results in shelling, debarring operations which prepare seed lots for basic seed cleaning
- ❖ It removes particles like pieces of trash, stones, clods, etc. that are larger in size than desirable crop seed from threshed seed lots
- ❖ Pre-cleaners in addition to remove big size particles, also removes particle that are light in weight and smaller in size than crop seed

Basic seed cleaning operation

- ❖ It shows actual cleaning and grading of seeds
- ❖ Just like pre-cleaning which may or may not be required, basic seed cleaning is essential in seed cleaning operations
- ❖ With this, different seeds can be cleaned and made into finished product

Upgrading quality of cleaned seed

- ❖ The various processing operations conducted after basic cleaning to improve seed quality are known as upgrading operations
- ❖ Upgrading operation depends on the type of contaminants and crop seed.

6.1.4 Grading

Introduction

- ❖ Market quality is identified if grade is established for certain crop.

- One of the objectives of grading seed is to ensure uniform quality, so that produce can be priced to satisfy the interests of producer and consumer.
- The grading standards employed and ability of handling or distributing to maintain standards till the produce is delivered to the customer.
- Seed is sorted so that seeds can be sown or in sets of predetermined number, decreases waste of sowing material and cutting down sharply on labour needs to tend plantings.
- Seeds can be graded or separated depending on their quality.
- Grading seed for next planting season is most important part of successful farming.
- Seeds can be graded as per emergence rates which they have, how fast a sapling grows from seed under optimum conditions of growth as big seeds posses 85% emergence rates and small seeds posses 72% emergence rates.
- Seed grading actively encourage consistency and competitiveness of seed as seed uniformity allows more even distribution of grain by the drill.
- Grading of seeds allows a grower to select only best quality seeds and once the grower have selected highest quality seeds, he/she can be best sure for good harvests.

In many countries, grading of seed depends on:

(1) Hectolitre mass

(ii) moisture content

(iii) foreign matter

(iv) damaged grains

Grading of seeds will give following marketing benefits:

- Low expenditure in transport and storage.
- Knowledge of prevailing price, and right markets.
- Easy financial assistance, and future trading.
- Widens the market for farm products.
- Consumers get wide choice of quality at a reasonable price.
- Promotes competitive marketing.

Grade Specifications:

Grading is undertaken by different agencies as per different criterias on usage. The bulk of commodity may be simply classified based on kernel texture such as:

(i) hard

(ii) semi hard

(iii) soft with colours

Need of grading

Grading of seed is required to:

- remove seeds of other crops which have been mixed unknowingly.

- remove weeds seeds, especially seeds which are resistant to herbicides and cannot be killed easily.
- remove soil dust, straw and other inert and harmful materials.
- remove immature, damaged, shrivelled, cracked, undersized and other seeds which are useless in nature.

Advantages of Grading

There are many advantages of grading such as:

- Grading before sale allow farmers to get higher price for their produce. It serves as an incentive to producers to market the produce of better quality.
- Grading facilitates marketing, for size, color, qualities and other grade designations of product that are well known to parties and requires no assurance from seller about product quality.
- Grading widens the market for the product, for buying which can be between the parties at distant places on telephone without any inspection of product quality.
- Grading reduces the cost of marketing by minimizing expenses on physical inspection of produce, thereby minimizing storage loses.
- Grading minimizing advertisement expenses and eliminating the cost of handling and weighing at ever stage.
- Grading makes it possible for farmers to get easy finance when commodities are stored, getting claims by railways/insurance companies, storage place for produce, market information, produce of different farmers, facilitate futures trading in commodity.
- Grading helps consumers to get standard quality products at fair prices.
- Grading contributes to market competition and pricing efficiency

6.1.5 Seed treatment

Introduction

Seed treatments control the pathogens that occur on or in the seed. The choice of seed treatment may be dictated by whether the pathogen is borne externally or internally. These pathogens may be important for three reasons:

- Some pathogens do not survive in soil or crop residue and are dependent on the seed borne phase for survival between crops such as fungus that causes loose smut of wheat.
- If a pathogen can survive in soil or residue, being seed borne may allow it to get a head start and, thus, result in more severe disease such as fungus that causes septoria leaf blotch of wheat.
- Seed borne pathogens may hitch a ride to new localities in seed shipments.

Description of seed treatment

Loose smut:

- Treat the seed with Tebuconazole @ 1g/kg seed or Vitavax @ 2 g/kg or Bavistin @ 2.5 g/kg seed for the control of loose smut.
- Seed treatment should be done with seed treating drum.
- Seed treatment should also be done with T. viride @ 4 g / Kg seed in combination with carboxin (Vitavax 75 WP) @ 1.25 g / Kg seed or Tebuconazole (Raxil 2 DS) @ 1.0 g / Kg seed.
- Seed treatment with T. viride alone @ 4 g / Kg seed is also helpful as it reduces the rust severity.

Flag smut:

- To control flag smut, treat the seed before sowing with Thiram @ 3 g/kg or Tebuconazole 1g/kg or Vitavax @ 2 g or Bavistin @ 2.5 g/kg seed.

Other diseases:

- Treat the seed with Captan or Thiram @ 3 g/kg seed for the control of root rot, foot rot, seedling blight, black tip and black spot of glumes.
- Captan and Thiram treatment should not be done earlier than one month of sowing as it affects seed germination.
- Treat seed before sowing with Trichoderma viride @4 g/kg seed.
- It provides induced systemic resistance, thus reducing the severity of the yellow rust at later growth stages of the crop
- In termite prone areas, seed treatment with chlorpyriphos @ 0.9g a.i or Endosulfan @ 2.4g a.i/kg seed, be taken up for their management.
- Seed treatment with thiamethoxam 70WS (Cruiser 70WS) @ 0.7 g a.i./kg seed or Fipronil (Regent 5FS @ 0.3 g a.i./kg seed) is also very effective
- The IPM module involves the seed treatment with T. viride (@4g/kg seed) + carboxin (75WP) @1.25g/kg seed) or Tebuconazole (@ 1.0g/kg seed) for the control of loose smut, followed by broadcast of insecticide treated soil (with endosulfan 35 EC @ 2.3L or chlorpyriphos @ 3L/ha) at 15 DAS for termites.

Ear cockle:

- The infested seed lot should be floated in 2 percent brine solution for this purpose
- The galls will float on the surface.
- These should be separated and destroyed away from the field by burning.
- The seed should be thoroughly washed to remove the salt solution before sowing.

Methods of seed treatment

- Due to infection in seeds, wheat crop gets different diseases like: smut, rust, blight etc. Due to these diseases, there is adverse effect on annual production of wheat.

- The infection from such diseases could be minimized or reduced by using only healthy, and quality seeds chosen through hot water seed treatment and treatment with fungicide, which is described detail below:

(A)Hot water seed treatment

- Initially, soak the seeds in ordinary water for about 4 hours, than dip in hot water 52 °C for 10 minutes or 55 °C for 5 minutes.
- After that, cool and dry the seeds in shade, and let for fungicide treatment.

(B) Fungicides treatment of seed

- It is compulsory to treat the seeds for producing quality and healthy wheat crop.
- Mix the wheat seeds before sowing with Bevistin or Benlate2 grams (0.2%) per kg seeds in an air tight pot and stir it vigorously for 8-10 minutes.
- Treating the seeds this way helps in controlling diseases and germs on outer surface of wheat grain such as blast, loose smut, etc.

(c) Seed Dressing

- The most common method of seed treatment.
- The seed is either dressed with a dry formulation or wet treated with a slurry or liquid formulation.
- Dressings are applied both on-farm or in specialised seed treatment facilities.

(D) Seed Coating

- A special binder is used with a formulation to enhance adherence to the seed and begin to impact seed size and shape.
- Coatings require advanced treatment application technology.

(E) Seed Pelleting

- The most sophisticated seed treatment technology, resulting in changing the physical shape of a seed to enhance plantability and handling.
- Pelleting requires specialised application machinery and techniques and is the most expensive of the applications.

Procedure and precautions for seed treatment

Precautions

The following things need to be considered while treating the seeds:

- Select non-windy place to treat seeds with fungicide so that air will not carry away the fungicide.
- Children, aged people, pregnant women are more vulnerable to poisonous effect of fungicide hence; they should not be involved in treating the seeds.
- Use mask, gloves, and goggles while treating seeds.
- Wear the clothes that cover whole body.

- Use only proper or prescribed amount of fungicides.
- Wash thoroughly with water when fungicide touches your skin directly.

Protection of Seeds and Seedlings

- Seeds and seedlings are vulnerable to many soil borne and foliar pests.
- Insects and pathogens can destroy germinating seeds and young plants, which are relatively tender and lack food reserves to recover from injuries or to survive extended periods of stress.
- Under stressful conditions, a number of aggressive or even fairly weak pathogens can become active and cause plant population and yield losses.
- Seed treatments can protect the seed and seedling from attack by certain insects and pathogens. No systemic fungicides or insecticides form a chemical barrier over the surface of the germinating seed which protect the germinating seed from chewing insects, such as wireworms, or soil borne pathogens, such as pythium.
- Certain systemic seed treatments can protect aboveground parts from sucking insects, such as aphids, or foliage diseases, such as rust.
- Systemic fungicides and biological seed treatments can also protect young plants from root rot. Although the duration of protection may be limited, a delay in infection can reduce the losses.
- Seed treatments lasts only for 10 to 14 days beyond planting, with pesticide breakdown being most rapid under warm, moist conditions.
- Certain active ingredients can protect seedlings considerably longer when applied at the highest labelled rate

Advantages

- Seed borne pathogens are vulnerable: The seed borne phase is weakest link in life cycle for many plant pathogens. With seed treatments to control seed borne pathogens is very effective for disease control.
- Precision targeting: Seed treatments are not subject to spray drift as chemicals are applied directly to seeds, little is wasted on no target sites like bare soil.
- Optimum timing: Seeds and seedlings are vulnerable to diseases and insects as compared to mature plants. Applying treatments to seeds will allow pesticides to be present when needed most.
- Low dose: Relatively small amounts of pesticides are used in seed treatments compared to broadcast sprays as it reduces the cost and the potential environmental impact and the probability of chemical residues in harvested grain.
- Easy to apply: Seed treatments are relatively easy and cheap to apply as compared to broadcast sprays.

Disadvantages

- Accidental poisoning: Treated seed looks like food to some animals as hungry livestock birds, such as pheasants or quail consume spilled treated seed.

- ❖ Cropping restrictions: Just like other pesticides, some seed treatments may have significant grazing or rotation crop restrictions.
- ❖ Limited dose capacity: The amount of pesticide that can be applied is limited by how much will actually stick to the seed. Seed coating technologies are helping to overcome this limitation, but phytotoxicity may still be a problem.
- ❖ Limited duration of protection: The duration of protection is short due to relative small amount of chemical applied to seed, dilution of chemical as plant grows, and breakdown of chemical.
- ❖ Limited shelf life of treated seed: Producing excess treated seed is undesirable as shelf life of treated seed is limited. Surplus treated seed cannot be sold for grain which is a serious limitation for seeds where seed germination and vigor decline relatively quickly.
- ❖ Phytotoxicity: Pesticide injury to plant tissues is called phytotoxicity as seed treatments exist in high concentrations on tender tissues of germinating seeds and seedlings having very low phytotoxicity.
- ❖ Worker exposure: In course of treating and handling large volumes of seed, workers may be exposed to seed treatment chemicals as aerosols. Inhalation of aerosols and skin contact with seed treatments must be prevented in the seed treatment process

6.1.6 Packing and bagging

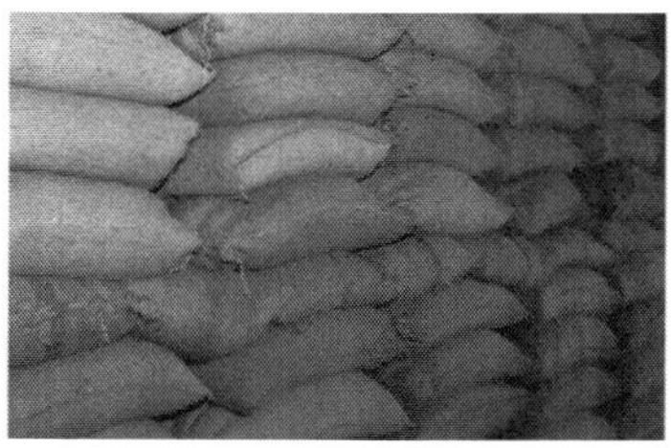

Fig Bags of seed

Proper packaging is an important element in reducing losses, especially in tropics, climate considerably increases the risks of seed deterioration. The main functions of packaging are to:

- ❖ permit easy handling, whether manual or mechanical;
- ❖ reduce product losses by theft;
- ❖ protect product from external attack (humidity, insects, sunlight, etc.).

There are various types of packaging for agricultural commodities, appropriate to product's characteristics and to marketing system. The choice of type of bag should also take into account not only its inherent toughness and resistance to humidity, sunlight and pests but also the type of handling anticipated.

Seed are normally stored in standard bags which are made either from jute or woven plastic. As per the size of storage, these bags are normally formed into stack. When using bag storage, consideration needs to be given to following:

- ❖ Jute bags should not be stacked higher than 4 m and plastic bags 3 m as plastic bags are more slippery and stacks will be less stable.
- ❖ Bags should be stacked under cover e.g. under a roof, in a shed or granary or under water proof tarpaulins.
- ❖ A one meter gap should be left between and around stacks and 1.5 m clearance between the top of the stack and the roof.
- ❖ Bags should be stacked on pallets or on an above ground structure to avoid the possibility of absorbing moisture from the floor.
- ❖ Bags should not be stacked on a bed of husks or bags filled with husks, as these are difficult to keep free from insect infestation.
- ❖ Bags should be stacked so that fumigation can be undertaken easily. The dimensions of the stacks should be set to facilitate sealing with a single fumigation sheet.
- ❖ The efficiency of bag storage can be improved if hermetic plastic liner bag is used inside the existing storage bag especially for seeds.

Storing of seeds bag

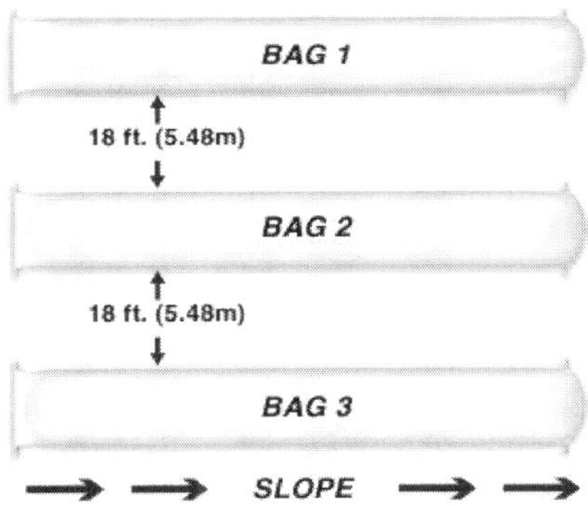

- ❖ The seed bags should be placed in north - south direction on storage site as it allow grain bags to be exposed to sun evenly.
- ❖ If seed bags are positioned in east - west direction, then south side of seed bags will be exposed to sun and may lead to overstretching and damage to seed bags.
- ❖ Keep the seed bag on highest elevation of storage site as it will aid in loading of seed bag and assist in preventing any moisture from entering the bag at the start and at finish points.
- ❖ Do not keep the seed bag across the slope as it could cause water build-up along the edge of bag which will add stress on one side of bag and may led to overstretch of bag.
- ❖ While storing seed bags side by side, it should be kept in mind that there should be proper space of around 18 ft. between two bags, as it will help for unloading equipment and trucks to travel between the bags.

- If the seed bags are stored in field, keep in mind that bags should be positioned in a line as it will help in preventing from animals, rodents and other pests.
- If the seed bags are stored in the depots, then position the bags in pairs with 5 ft. distance apart or 18 ft. in case of pairs.

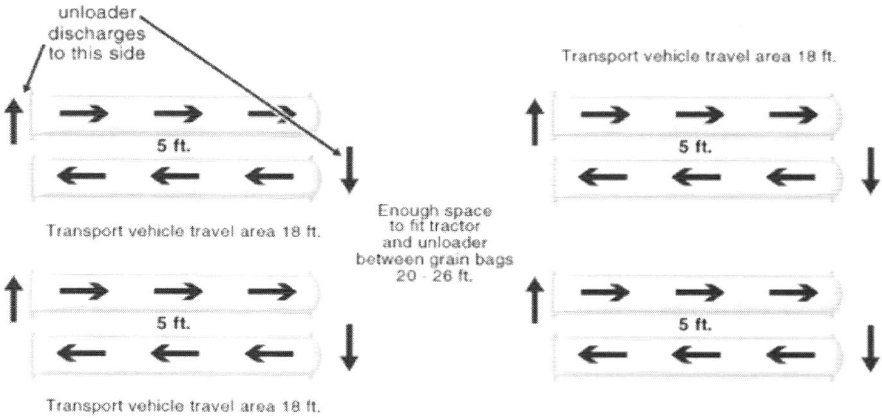

Precautions

For well-managed reception and storage of seed delivered or stored in bags, following rules should be observed:

- avoid taking in more seed than storehouse can hold as capacity depends on number of individual lots one wishes to store;
- store only seed that are very dry and clean;
- recondition products in case of damp or torn bags, or if the quality of seed seems doubtful;
- erect stacks of bags that are stable and easily accessible;
- maintain the cleanliness and good condition of premises and surroundings, equipment, and stored products;
- make timely provision for supplies of fuel, bags, insecticides and miscellaneous products.

6.2 Ideal seed storage, different factors influence the storage, type of storage structures, warehouse and different conventional and advanced seed storage structures.

6.2.1 Seed Storage

- Proper storage of seed is important to conserving its vigor and vitality.
- Seeds can last from few years to centuries, as per its species and storage conditions.
- Many seeds are stored in cool, dry conditions which will survive for longer as compared to seeds that are stored in wet, warm environment.

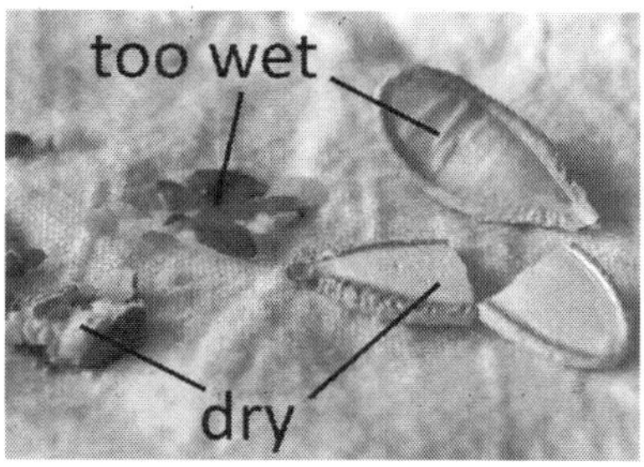

- When the seed matures on plant, slow process of deterioration begins, so soon that seeds should be kept in store for betterment.
- Seeds must be stored in a way which maintains their viability for long periods.
- The seed banks that store seeds to preserve genetic diversity, where seeds are dried to optimum moisture content, which are evaluated for quality and genetic purity and are sealed in moisture-proof containers.
- Seeds left at ambient temperatures and relative humidity's will lose their viability quickly whilst seeds stored in conditions of low moisture content and temperature will retain their viability for longer periods.

Loss of seed viability in storage

There are certain conditions that cause loss of seed viability in storage which include:

- immature or damaged seed will not survive longer storage periods, so seed should be harvested when properly matured;
- mechanical injury to seed during harvest or handling makes it more susceptible to deterioration in storage;
- seed should be properly dried before going into storage and protected from moisture and high relative humidity, as fungi causes damage to stored seed if seed moisture is high;
- high storage temperature has damaging effect on seed, so stores should be designed so that low temperatures are maintained;
- some seed treatments cause seed to die if it is stored too long, so seed should only be treated when it is certain that it will be sold for planting;
- rodents like rats and mice, can be most destructive to seed, so effective rodent control is required in all seed stores;
- insects to be controlled by insecticides and fumigants, so grower needs to use safe fumigants as many fumigants can reduce germination.

6.2.2 Different factors influencing seed storage

Introduction

- The stage of maturity at harvest will affect seed viability and longevity.

- Seed that is too wet for immediate storage can be slowly dried with warm air, bringing the moisture down to 15% before storage.
- Care must be taken not to use air temperatures above 108^0 F to dry the seed, as it will cause reduced viability or death of seed.
- To maintain viability, native seed should be stored in bulk at 10-12% moisture and in bags at 12- 15% moisture.
- Seed with long dormancy period, may require a period in storage of 14-18 months in order to completely break dormancy.
- When environmental conditions are not controlled during storage, native plants will reach their maximum germination 2-3 years following the harvest and maintain high level of viability for 3 to 6 years.
- Seed should be stored under carefully controlled environmental conditions to maintain viability over long periods.
- To maintain existing viability, seed has to be stored cool and with low relative humidity.
- For low humidity/moisture, sum of air temperature and relative humidity should be less than 100 to provide suitable long-term storage environment.

Principles of Storage

- Seed storage requires cool temperatures, low humidity and dim to no light.
- Seed storage conditions should be dry and cool
- Effective storage pest control
- Proper sanitation in seed stores
- Before placing seeds into storage they should be dried to safe moisture limits.
- Storing of high quality seed only i.e., well cleaned treated as well as of high germination and vigour.

Genetic factors

- The storage is influenced by kind/variety of seeds.
- Some kinds are naturally short lived, as within a crop, storage period varies between varieties.

1. Pre harvest factors

- Effect of provenance: Seeds produced can be stored for longer period due to different climatic conditions and soil types.
- Effect of weather: Fluctuating temperature during seed formation and maturity, affects seed storage.
- Pre harvest sanitation spray: In pulses, insect infestation comes from field (e.g.) bruchids.
- Seed structures

- Quality of seeds: Seed lot with vigorous, undeteriorated seeds will be stored for longer as compared to deteriorated lots.

2. Environmental factors

- Moisture content: Amount of moisture in seeds is important which influences seed viability during storage. If seed moisture content increases, storage life decreases while if seeds are kept at high moisture content, losses could be very rapid due to mould growth very low moisture content.
- Relative humidity and temperature during storage: Relative Humidity and temperature are important factors in determining storage life of seeds. If there results high RH in increase in seed moisture contentment, it results in faster deterioration of seed quality. Low humidity will help seeds for longer storage. Equilibrium moisture content for particular seed at given relative humidity tends to increase as temperature decreases.
- Temperature: Temperature is imporant in life of seed. Insects and moulds increases with increase of temperature. Higher the moisture content of seeds, more will be the adverse affect. Decreasing temperature and seed moisture is good means of maintaining seed quality in storage.

6.2.3 Types of seed storage structures

6.3 Perform fumigation and seed treatments in seed stores

Fumigation is a method of pest control which fills an area with gaseous pesticides to suffocate or poison the pests. It is used to control pests in buildings, soil, grain, and is used during processing of goods to be imported or exported to prevent transfer of exotic organisms.

Properly conducted fumigation will stop insect infestation and grain degradation from getting progressively worse. When fumigation is effectively conducted in late fall, pest populations can be drastically reduced.

Metallic drums or wooden boxes are used for fumigating small quantities of grain. Ethylene dichloride and carbon tetrachloride mixtures are recommended for fumigation of food grains in storage at farm level while hydrogen phosphide in form of aluminium phosphide or methyl bromide for protection in warehouses, godowns and silos.

Mixture of ethylene dichloride and carbon tetrachloride @ 1 litre for 20 quintals of grain or 35 litres per 100 m^3 of space with exposure period of 4 days is recommended. Methyl bromide is used at the rate of 3.5 kg per 100 m^3 of space with 10-12 hours exposure. The fumigant, hydrogen phosphide (aluminium phosphide), is available in tablet form and can be used at the rate of one tablet (3 g) per metric tonne or 25 tablets per 100 m^3 of space with an exposure period of 7 days.

For successful fumigation, following factors needs to be considered:

- ❖ Grains should always be level in store to allow fumigant to penetrate evenly.
- ❖ Any surface caking or crusting should be broken up and removed.
- ❖ Grain temperature should be 60°F or higher to ensure proper vaporization.
- ❖ Possible leak points like cracks or holes in store should be closed before fumigation as leakage will led to under treatment and poor control.

6.4 Understand storage pests and diseases and their management

6.4.1 Introduction

Food stores are good place for breeding of pests. A farmer/seed grower has to take adequate measures about seed storage to conserve the crops to be able to feed the family even during droughts. The main storage pests, apart from rodents, are beetles and moths. Some pests like grain borers, weevils and Angoumois grain moths are able to feed on whole, healthy seed/grains, and are considered as primary pests while flour beetles that attack only broken grain, moist and thus soft grain are secondary pests.

6.4.2 Insects pests of seeds

Greater False Wireworm

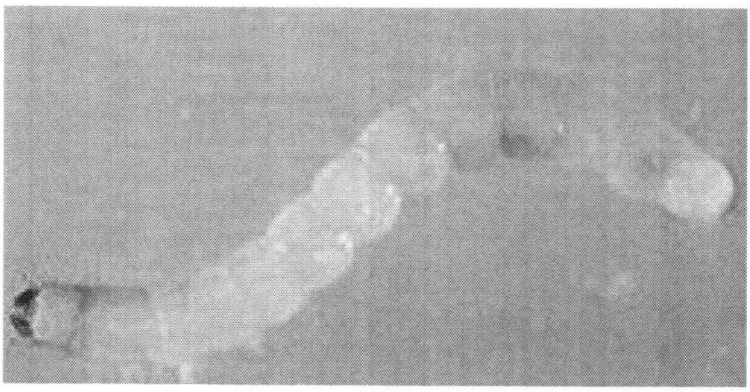

- ❖ It is the larval stage of dark-colored beetles with long legs seen on soil surface and under plant litter.
- ❖ In this, larvae feed on the seed and stems of seedlings below soil surface which causes stem to have tattered appearance.
- ❖ The adult beetles damages emerging seedlings which eventually wilt and die.

Lesser False Wireworm

- It occurs sporadically throughout summer rainfall region.
- It is the larval stage of dark coloured beetle 'Dusty Surface Beetle'.
- The larvae feeds on seed and stems of seedlings below the soil surface which causes stem with tattered appearance.
- It makes the plants wilt and die leading to seedling density-loss in field.

Black Maize Beetle

- It occurs sporadically throughout the summer rainfall region.
- The adult beetles have shiny black colour and are 12 - 15 mm in length.
- They are found in or on the soil surface layer.
- It feeds on the seedling stems at or just below the soil surface.
- It lead to big holes with ragged edges after feeding which results in plant death and seedling density-loss.
- Its larvae feed on decaying organic material and do not damage the crop.

Western Province Grain Worm

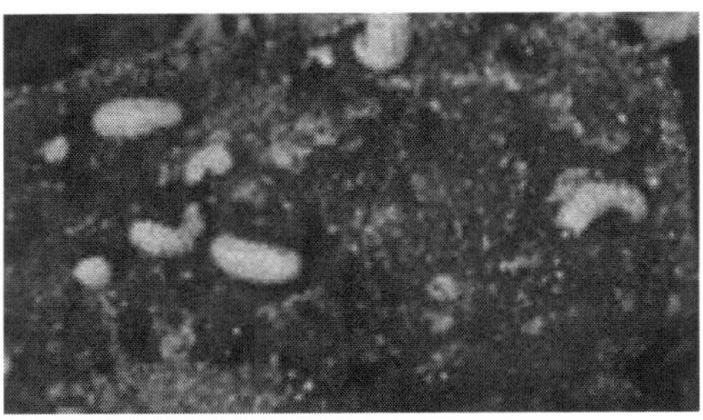

- It is a sporadic and unpredictable pest of wheat under intense rainfall conditions.

- In this, C-shaped larvae are legless and can reach 10 mm in length.
- It exist in white to cream coloured with small shiny brown heads.
- Its larvae feed on the stems of seedlings below the soil surface which causes stem to have tattered appearance.
- It makes the plant to wilt and die leading to poor plant standing in fields.
- The adult beetles feed on the leaves and will not economically damage the plant.

Sandveld Grain Worm

- It is 10 - 12 mm in length.
- The adult beetles are blue-green to metallic green and is 6 mm long.
- In this, the pupae are present in cells below the soil surface.
- Its larva is of cream colour with dark brown head is feed on stem of plant below the soil surface which lead small plants to die.
- Big plants survive with their feeding damage and tillering of plants gets reduced.

6.4.3 Symptoms of seedling disease

Symptoms of seedling disease appear as patchy emergence during 4 weeks following seeding, or up to 4-leaf stage. Patchy emergence results from wide range of factors such as insects and soil conditions. There are following seed disease symptoms such as:

- Seeds fail to germinate and become soft and pulpy resulting as seed decay.
- Seeds germinate but developing seedlings decay and fail to emerge is pre-emergence damping-off.
- Seedlings emerge and appear normal above ground, but either roots decay with rot moving rapidly up in hypocotyl, or young stem may be partially or completely girdled with decay and shrivel at any point where young above-ground parts of canola seedling exhibits purple or chlorotic discoloration.
- If decay reaches soil surface, emerged part of seedling topples over, wilts and dies known as seedling blight, wirestem or post-emergence damping-off.
- The hypocotyl appears constricted or shriveled and discolored reddish brown.
- In moist topsoil, shriveled stem persist for while, while in dry windy conditions, whole seedling disappears in few days.
- Seedlings emerge and stagnate in 2-4 leaf growth stage even when growing conditions appear favorable where rootlets appear to be missing, known as seedling root rot.

6.4.4 Seed Diseases

Seed diseases are caused by fungi which may be on seed during planting or by fungi in soil and are responsible for poor emergence and thin stands. The fungi grow in soil when conditions are suitable or are stimulated by secretions from germinating seeds or roots of host plants. The fungi are weak pathogens which will able to infect only young succulent roots and young stem tissue. Plants with damaged root systems may stagnate for a while, then grow and mature in the usual way.

Once in plant, fungi multiply, it causes decay that damages or kills the seedling. The fungi can form microscopic resting bodies within or on the surface of the infected tissue following the death of the cells. The resting or dormant bodies allow the fungi to survive until another susceptible host becomes available. At two- to four-leaf stage, below ground parts of plants become sufficiently woody to withstand further infections and are able to regenerate more rootlets than they lose. The fungi may persist throughout the life of the plant on its feeding rootlets, reducing the efficiency of the root system in utilizing moisture and nutrients, and reducing tolerance to moisture stress in midsummer heat which results in stand with poor vigour and reduced yield.

Seedling diseases take place after germination till shortly after emergence. If disease occurs before emergence, then seedling may rot before it has a chance to get out of soil and such is referred to as pre-emergence damping-off. After emergence, seedling stems is attacked at or below the soil line, which is known as post-emergence damping-off. There are two general symptoms which occur during this time, brown to reddish-brown or black cankers at soil line and soft watery rot. In first instance, if cankers girdle the stem, seedling falls over and dies; or, with partial girdling, the plant may continue to live, but will be stunted. In other instance, soft watery rot continues till entire seedling decays. Seed and seedling diseases are most common in cool, wet soils which are controlled by planting crops in warm soil, by fungicide seed treatments and using in-furrow, broadcast or band-applied fungicides.

Damping-off and seedling blight occurs before seedling emergence (pre-emergence) or after seedling emergence (post-emergence). Pythium incited damping-off is characterized by high order of pre-emergence killing of seedlings where deterioration process begins soon after the seed coat is broken.

In post-emergence damping-off, seedlings emerge above the soil and begin deteriorating at soil level. As deterioration progresses upward, seedlings appear water soaked. As tissue collapses, the seedlings shrivel and turn brown. Pythium, Fusarium, Rhizoctonia, Helminthosporium and Curvularia all cause post-emergence damping-off. The effect of such seedling diseases is significantly reduced stand of grass which usually needs replanting. If disease outbreak is treated in time, stand may only be thinned by the attack.

Control

- To control the incidence of seedling diseases, grower should use only top quality seed, as seed that germinate quickly and produce vigorous plants can survive seedling stage.
- Grower should plant the seed at proper time, as planting seeds too early also increases risk of seedling diseases.
- As per location, grower should plant cool season grasses from mid-September to November as good temperatures for planting cool season grasses occur when night temperatures are in mid-60's and day temperatures is in mid-80's.
- Grower should make an application of soluble nitrogen fertilizer several days before planting of seeds to ensure availability of nitrogen.
- Grower should not overwater the seedbed.

- ❖ Preemergence damping-off by Pythium be reduced by using seed treated with fungicide.
- ❖ Treating seedbed with broad spectrum fungicide should be done before seeding reduces incidence of pre-emergence seedling losses.
- ❖ After seedling emergence, fungicide applications can control the damping-off, fading out or seedling blight.
- ❖ Fungicide applications between broad spectrum fungicide and Pythium specific fungicide at 7-10 day intervals during seedling stage.
- ❖ During seedling stage, grower should keep soil moisture sufficient to avoid drought stress.
- ❖ Grower should avoid heavy applications of soluble nitrogen and mow with sharp mower to avoid pulling up young seedlings.
- ❖ Grower should use syringing to wash dew off foliage in the marning to reduce the incidence of seedling diseases.

6.5 Practice seed quality enhancement techniques- priming, coating, pelleting etc.

6.5.1 Information

Seed enhancements are technologies that are gaining increasing attention for potential to confer greater disease resistance in seeds, improves seed vigor and modify seed emergence capabilities. For any crop, time from sowing to seedling establishment is crucial where seeds are exposed to wide range of environmental stresses. Seed quality is main to sustainable crop production and food security. Seed enhancements include physical, physiological and biological treatments to overcome germination constraints by uniform stands, earlier crop development and better yields.

6.5.2 Objectives of Seed quality enhancement

- ❖ Enhance germination/seedling growth by manipulation of seed vigour or physiological status of seed.
- ❖ Facilitate seed planting.
- ❖ Deliver the materials required at sowing (e.g. nutrients, inoculants).
- ❖ Remove weak or dead seeds using non-traditional upgrading techniques such as density, color, sorting.
- ❖ Tag seeds with visible pigments or other materials/markers for traceability and identity preservation.

6.5.3 Seed Priming:

- ❖ It is the process of controlled hydration of seeds to level that allows pre germinative metabolic activity to proceed, thereby preventing actual emergence of radical.
- ❖ It is used to increase rate of germination and uniformity.
- ❖ In this, seeds are dried, distributed and planted in normal way.

- Primed seed shows more rapid and uniform emergence of seedlings from soil as compared to non-primed seed of same seed lot.
- Primed seed has shorter shelf life than non-primed seed, and are stored under optimal conditions before planting.

Types of Seed Priming

- Osmo priming: Soaking the seeds in osmotic solutions.
- Halo priming: soaking the seeds in salt solutions
- Bio priming: soaking the seeds with biological agent's solution like Bacteria, Rhizobium etc.
- Solid priming: This consists of mixing seeds in an organic or inorganic carriers and water for a particular period of time. The moisture content of the matric brought to a level just below what is required for radicle protrusion. Seed water potential is regulated by the matric potential of the seed and during priming water is held by carrier of seeds can imbibe water form carrier till the equilibrium is reached.
- Matrix priming: mixing with moist solid particulate materials, such as exfoliated vermiculite, diatomaceous earth or lignaceous shale.
- Hydropriming: controlled imbibition, i.e. the continuous or staged addition of a limited amount of water, such as in 'drum priming', though hydropriming is also used to mean imbibition in effectively unlimited water for a short period of time.

Priming Agents

- Water
- Salts, NaCl, Na_2SO_4, KNO_3, $CaCl_2(NH_4)SO_4$, KCl etc
- GRs: GA_3, CCC, Kinetin
- Vitamins: Vitamin K_3, Nicotinic acid, Panthonic acid
- Plant products: Garlic extract, Coconut water, Leaf extract of Pongamia pinnata, Abizzia amaza, Prosopis juliflora.

6.5.4 Seed Coating

- Seed coating is defined as substance applied to seed that does not obscure its shape.
- It includes fungicides, microbiological treatments and micronutrients.
- In this, seed enhancement material is kept on to the seed.
- In this, active ingredient is applied in quick-drying polymer film around the seed, for gaining popularity.
- It reduces loss of active material from seed during seed transport and handling.
- In this, small amount of seed is required as compared to broadcasting, which is most economical for improving seed performance.

Methods of Seed Coating

The seeds are usually coated by the following methods:

a) Dry seed dressing:

- Seeds are treated with active ingredient or coating materials along with adhesives such as liquid form.
- The seeds are stirred well or rotated in a drum for uniform coating.
- The coated seeds are shade dried and used for sowing.

b) Slurry seed dressing:

- In this, solvents or diluents like water or fevicol is used.
- The concentration of diluents varies greatly with the seeds.
- The seeds are coated with active ingredients, simultaneously with adhesives or polymers diluted in diluents or solvents.
- After using diluents, seeds are stirred uniformly and coated seeds are allowed to shade dry and used for sowing.
- Seed grower should take care for use of optimum concentration of adhesives and diluents for having effective coating.

6.5.5 Seed pelleting

- Pelleting is coating of seeds with inert materials to make uniformity in size and shape.
- It is a technique for increasing the size of small seeds, and encapsulating them to make bigger, rounder, smoother, and more uniformly-sized seeds.
- It is a coating technique used for handling small seeded crops and increases precision planting.
- It improves seed quality and reduces seed rate and deterioration during storage in case of vegetable crops.
- Pelleted seed could be direct seeded in field/sown in flats for transplants which allows more accuracy and results in an efficient seeding process.
- Seed pelleting have been developed for seed signalation, adding bioactive chemicals, nutrients and beneficial microbes for sowing seed under dry and wet conditions.

Unit 7

Seed Quality Testing

7.1 Concept of seed quality

7.1.1 Seed Quality

Seed quality is the possession of seed with required genetic and physical purity which carries physiological soundness and health status. Seed quality is perhaps the most important aspect of seeds production, acquisition and use. A bad seed or seed lot is not only a waste of money, it is a waste of time and it brings mistrust among farmers, seed producers and seed sellers. Bad quality seeds affect the entire crop production value chain. Even more important is that bad quality seed with low germination power or count can discourage farmers from purchasing them especially where efforts are made for introducing improved seed varieties.

There are four basic attributes of seed quality to consider:

- Physical qualities of the seed in a specific seed lot
- Physiological qualities observes germination and vigour of seeds
- Genetic quality which relates to specific genetic characteristics of seed variety
- Seed Health which refers to the presence of diseases and pests within a seed lot.

Physical Quality

- It is the cleanliness of seed from other seeds, debris, inert matter, diseased seed and insect damaged seed.
- The seed with physical quality should be of uniform size, weight, and colour.
- The seed should be free from stones, debris, and dust, leafs, twigs, stems, flowers, fruit well without other crop seeds and inert material.
- Seed should be devoid of shriveled, diseased mottled, moulded, discoloured, damaged and empty seeds.
- The seed should be easily identifiable as species of specific category of specific species.
- Lack of quality character in seed will indirectly influence the field establishment and planting value of seed.

- The quality character of seed could be obtained with seed lots by proper cleaning and grading of seed after collection and before sowing / storage.

Genetic purity

- It is true to type nature of seed.
- The seedling / plant / tree from the seed should resemble its mother in all aspects.
- The quality character of seed is important for achieving desired goal of raising the crop either yield or for resistance or for desired quality factors.

Physiological Quality

- It is the actual expression of seed in further generation / multiplication.
- Physiological quality characters of seed have seed germination and seed vigour.
- The liveliness of seed is known as viability.
- The extent of liveliness for production of good seedling or ability of seed for seedling with normal root and shoot under favorable condition is known as germinability.
- Seed vigour is energy or stamina of seed in producing elite seedling which is the sum of all seed attributes that enables its regeneration under any given conditions.
- Seed vigour shows level of performance of seed or seed lot at germination and seedling emergence.
- Seed that do well at sowing are termed as quality seed and based on degree of performance in production of elite seedling, it is categorised as high, medium and low vigour seed.
- The difference in seed vigour is differential manifestation of deteriorative process which oocurs in seed before ultimate loss of ability to germinate.
- Difference in seed vigour is expressed in rate of emergence, uniformity of emergence and loss of seed germination.
- The viable seeds needs no germination but all germinable seed will be viable.
- The vigourous seeds will be germinable but all germinable seed need not be vigourous. Physiological quality of seed is achieved by proper selection of seeds used for sowing and by caring for quality characters during extraction, drying and storage.
- Seed with good vigour is preferable for raising a good plantation as the fruits, the economic come out are to be realized after several years.
- The selection of seed based on seed vigour is important for raising perfect finalize plantation.

Seed Health

- The health status of seed is the absence of insect infestation and fungal infection, in or on the seed.
- Seed should not be infected with fungi or infested with insect pests as these will reduce the physiological quality of seed and also the physical quality of the seed in long term storage.

- The health status of seed includes deterioration status of seed expressed by low vigour status of seed.
- The health status of seed influences the seed quality characters directly and warrants their soundness in seed for production of elite seedlings at nursery / field.

7.2 Seed testing and its role in crop management

7.2.1 Seed Testing

Seed testing is the science of evaluating the quality of seeds to determine their value for planting. Though initially developed for field and garden seeds, seed testing is equally valuable for determining the seed quality of turf, flowers, herbs, shrubs, trees and native species.

Seed testing measures the viability and physical factors which regulate use and maintenance of seeds.

7.2.2 Objectives

The idea of seed testing is to:

- determine seed quality which can be suitable for plantation.
- identify seed quality problems and their causes.
- find the need for drying, processing and procedures that needs to be used.
- find if seed meets required quality standards or labelling specifications.
- establish quality
- provide basis for price and consumer discrimination among lots in the market.

7.2.3 Components

The components of seed testing involve:

- Seed physical purity
- Germination
- Seed health
- Seed vigour testing
- Seed sampling
- Seed moisture
- Seed viability

Seed physical purity

- The idea of testing seed purity is to determine the composition by weight of sample being tested, and by inference the composition of seed lot and identity of various species of seeds and inert particles in sample
- The purity test involves an examination of working sample which can be 1/10th of the size of submitted sample.

- ❖ Each working sample is divided into pure seeds, other seeds and inert matter.
- ❖ In this, each separation is weighed and expressed as percentage of whole working sample weight to give pure percentage of seed, other crop seed, weed seed and inert matter.
- ❖ The balance of submitted sample is examined for other contaminant species which are available in purity test.
- ❖ This test is not done on regular basis and is not required for international analysis certificate till specially requested.

Germination

- ❖ In lab test, germination is emergence and development from seed embryo of essential structures which for kind of seed being tested shows its ability to develop in normal plant under favourable conditions in soil.
- ❖ It involves keeping four replicates of 100 pure seeds on damp blotting paper in germination cabinet at suitable temperature for rapid germination.
- ❖ In this, germination counts are made after specified incubation periods, which depend on species.
- ❖ Here, white clover counts are made after 3-7 days.
- ❖ In case of legume seeds, extra assessment of hard seeds is made at final count.
- ❖ With certain types of seeds, many grass seeds tested soon after harvest, special treatments like pre-chilling for 4 days is done at 5°C or use of blotters soaked in potassium nitrate, to be used to overcome postharvest dormancy.
- ❖ The use of such treatments is shown, but level of post-harvest dormancy is not routinely reported as it disappears before sowing.
- ❖ Due to certain doubts expressed about relevance of germination testing, such technique is generally most reliable in assessing plant value of seed lots.

Seed health

- ❖ Number of diseases affect pasture seed production that are transmitted by seed.
- ❖ Seed health testing at station is not done on samples of pasture seed, except where fixed need has been established.
- ❖ As seen, samples of ryegrass seed were collected before harvest, that were examined at station for presence of blind-seed disease infection.
- ❖ In this, if crops were found to be severely infected, they were made into hay, thereby avoiding harvesting and cleaning costs for product which was worthless.
- ❖ Many seed-borne diseases of pastures have been subject to periodic surveys, while it has not been considered necessary to institute any routine seed health testing.
- ❖ Certain serious seed-borne diseases linked with seed production of new cultivars of prairie grass and paspalum that needs control by routine seed health testing and seed certification scheme.

Seed vigour testing

- It refers to certain properties of seed which influences speed and uniformity of germination and ability to germinate and emerge under broader range of field conditions.
- Lack of seed vigour results as slow or uneven field emergence, or failure to establish and laboratory germination tests if indicated which effects leading to reduced crop yields.
- The effects can be altered by sowing conditions like soil type, soil temperature and availability of soil moisture.
- Differences in ability to retain germination during storage attributes to differences in seed vigour.
- Many types of laboratory test have developed for seed vigour determination especially for large-seeded crop species.
- No single test satisfies all needs and method/combination of methods selected to suit species or cultivar concerned and environment to which, seed likely to be sown.
- The tests having broader application are difficult to develop and are not in routine use.
- There are many causes of low seed vigour that are controlled by best possible conditions for growth of parent crop, proper handling and storage facilities for harvested seed.

Seed moisture

- Moisture test determines the overall moisture of seedlot to establish optimal conditions for storage and to maximize lifespan of seed.
- In this, testing methods are not interchangeable, i.e. eRH reading should not be converted to MC for test result submissions and vice versa.
- Moisture testing should normally be done within 24 hours of receipt at testing facility.

7.2.4 General quality standards

Physiological stage

- For quick emergence and a good growth vigour, seed should have been stored for at least 2.5-3 months and no longer than 5-9 months as per storage temperature.

Virus diseases

- Seed to be used for multiplication should not contain more than 2% virus, so during field inspection, tolerance level of 1% virus is used.
- Seed supplied to higher altitudes to farmers organizations which produces own seed or seed for the market, the tolerance for virus diseases should not be more than 2% virus.
- Seeds to be used for production of table potatoes, more tolerance levels should be applied.

- A virus percentage of 10% gives yield reduction of not more than 2-3%.

Earth and extraneous matter

- Not more than 2% by weight.

Dry and wet rot

- Dry and wet rot which includes Phytophthora infestans has around 1% weight, except if caused by Synchytrium endobioticum, Corynebacterium sepedonicum, Pseudomonas solanacearum.

Tubermoth

- Not more than 2% of the tubers attacked.

Common scab

- Over 25% of tuber surface affects 5%.

Rhizoctonia

- Over 10% of tuber surface affects 5%

7.3 Understand seed sampling in seed stores and labs

7.3.1 Seed Sampling

- The idea of sampling is to obtain a sample that is good in size for testing and to obtain sample, that is representative of lot being tested.
- The sampling should be done quickly with limited exposure to room air to minimize changes in seed moisture if to be tested.
- Seed testers reminds that any movement and settling of seed in containers can bring empty lighter seeds to surface, so mixing before any samples are taken to ensure accurate representation of seedlot.
- Primary samples are small part of seed taken at random from seedlot.
- The primary samples should be taken from one seedlot, which are combined and mixed to form composite sample.
- The composite samples get reduced to smaller subsample called as submitted sample.
- The submitted samples are part of seed that are submitted to lab for testing.
- Working samples are subsamples of submitted sample and is the part of seed on which a test is to be made.

7.3.2 Sampling intensities

The following sampling intensities shall be used:

- 1-4 containers = 3 primary samples from each container
- 5-8 containers = 2 primary samples from each container
- 9-15 containers = 1 primary sample from each container

- 16-30 containers = 15 primary samples from lot taken from randomly selected containers with not more than one sample per container
- 31-59 containers = 20 primary samples from lot taken from randomly selected containers with not more than one sample per container

Primary samples

- The method for obtaining primary samples must meet certain standards.
- The use of trier/sampling stick or sampling by hand to have cheap but effective sampling.
- Primary samples taken by hand to be taken from middle of container if one sample is taken, top & bottom if 2 samples are taken, top/middle/bottom if 3 samples are taken, etc. to ensure representative composite sample.

Composite sample

- The primary samples are combined and mixed to make uniform composite sample.
- A random method to be used to reduce composite sample to required submitted sample size.
- The submitted sample to be transported in an intact, hermetically sealed container with little air as possible.
- Plastic Zip bags are not recommended and are only acceptable for very short storage periods of less than an hour.
- Glass vials, rubber sealed glass jars or heat sealed foil bags are recommended as containers which are hermetically sealed at room temperatures, with heat sealed foil bags being most reliable and cost effective.
- The seed grower/tester should start the germination tests within 2 weeks of receiving the submitted sample.
- Interim storage should be in refrigerated conditions.
- Moisture tests must be started within 24 hours of receipt of submitted sample, but samples should be brought to room temperature before containers are opened.

7.4 Determination of genetic purity in lab and field.

7.4.1 Introduction

- Genetic purity is a quality criteria required for successful seed production
- Genetic purity is true to type plants/seeds that conforms to characteristics of variety as shown by breeders.
- Genetic purity or genuiness of cultivar is tested by using heritable characters of seeds, seedlings or plants.
- The genetic purity testing involves use of computerized systems to capture and process morphological information and to use of biochemical methods to analyze many components of seeds.

- The methods used in assessing genetic purity involves morphological/conventional grow out test, chemical test, electrophoresis method, biochemical markers, molecular markers.

7.4.2 Seed physical purity

- The idea of testing seed purity is to determine the composition by weight of sample being tested, and by inference the composition of seed lot and identity of various species of seeds and inert particles in sample
- The purity test involves an examination of working sample which can be 1/10th of the size of submitted sample.
- Each working sample is divided into pure seeds, other seeds and inert matter.
- In this, each separation is weighed and expressed as percentage of whole working sample weight to give pure percentage of seed, other crop seed, weed seed and inert matter.
- The balance of submitted sample is examined for other contaminant species which are available in purity test.
- This test is not done on regular basis and is not required for international analysis certificate till specially requested.

7.5 Estimation of seed viability and vigour in different types of seeds

Seed viability

- It is an ability of embryo to germinate which is affected by many conditions.
- There are many factors that affects seed viability such as ability of plant to produce viable seeds, predator and pathogen damage, and environmental conditions.
- The age of seed also affects its health and germination ability.
- As seeds are living embryos, where overtime cells die, that cannot be replaced.
- The amount of time a seed remains viable can be influenced by genetics and environment.
- Many seeds remain viable under optimal conditions for many years, while others for season cycle.
- Seed viability is important for industries like forestry and agriculture, as they depends on seed germination.
- In seed viability, seeds can be predicted based on visual inspection of seed quality and germination of sample population of seeds.

Seed vigor

- It is the total of properties of seed that shows the level of activity and performance of seed/seed lot at the time of germination and seedling emergency.
- In seed lot, losses of seed vigour are related to reduction in ability of seeds to carry physiological functions which allows them to perform.

- In this, physiological ageing starts before harvest and continues during harvest, processing and storage.
- It progressively reduces performance capabilities as a result of changes in cell membrane integrity, enzyme activity and protein synthesis.
- The biochemical changes which occur quickly/slowly depends on genetic production and environmental factors.
- The end point of deterioration results in death of seed.
- The seeds lose vigour before they lose ability to germinate, so seed lots having similar high germination values can differ in physiological age and in seed vigour.
- The seed vigour differences exist in seed lots of agricultural, horticultural and silvicultural species.

Seed Vigor Testing

- Seed vigor testing provides precise identification of important differences in physiological potential among seed lots of commercial value, with similar germination percentage, to identify lots of higher probability to perform well after sowing and/or during storage.
- In this, need of efficient vigor test has high sensitivity to differences in physiological potential not detected by viability tests and ability to rank lots as per performance potential.
- It is require to find high and low vigor seed lots in comparable way to field seedling emergence.
- From the figure, consistent evaluation of seed physiological potential involves use of standardized procedures and corrects interpretation of meaning and scope of results.
- In vigor testing, seed physiological quality is measured which is understandable and natural where evaluation of seed vigor needs more rigorous control of test variables and interpretation criteria to get consistent results.

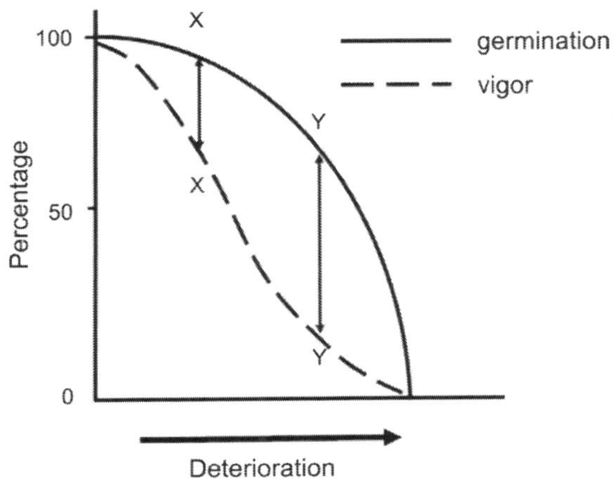

- In the figure, technologists and consumers understand that vigor tests were not designed to predict the exact number of seedlings that will emerge in a stand in field or time a seed lot will be able to maintain "x" percentage germination during storage for "n" months, so proper interpretation of test results shows information for selection of better seed lots, determination of the required quantity of seeds and distribution during sowing, and estimation of probable commercial success under various field conditions.
- Seed vigor is combination of characteristics that shows potential for high performance after sowing.
- There are many techniques to determine its assessment, such as those that are directly or indirectly evaluating current seed metabolic state to establish relationship with seedling emergence and storability, which covers tests such as electrical conductivity, tetrazolium, and tests that evaluate seedling growth.
- Many tests are done with an objective of finding seed tolerance to stress, as more important stress tests has cold test, accelerated aging, and controlled deterioration.

7.6 Seed health testing for seed borne diseases and insect pests

7.6.1 Introduction

- Seed testing is done to establish whether seed is infected or not.
- Seed health tests detect the most important seed-borne pathogens in crops.
- The seeds which is pathogen-free should be used for sowing, as testing of seed before sowing will show potential disease problems which allow steps that reduces disease risk.
- Lab testing is usually required, as infected seed may have no visible disease symptoms.
- The results of seed health tests shows serious diseases of pulses which are seed-borne and significant crop losses that result from usage of infected seed.
- The uncontrolled movement of infected seed between regions result in rapid expansion of area affected by such diseases.
- The pathogens can adversely affect germination, cause seedling infection and damage mature plants.
- The transmission of fungal and bacterial pathogens from seed to crop can vary considerably depending on growing conditions. Diseases caused by viruses usually have higher transmission rates than those caused by fungi or bacteria, and are less affected by seasonal conditions.
- Seed-borne diseases in early growth of plant causes poor crop establishment and reduced plant vigour, resulting in low yields.

Seed borne inoculum may:

- Cause disease in the crop
- Introduce disease to new regions
- Reduce germination and % normal seedlings

7.6.2 Seed Health Testing

- Seed health testing is an essential management tool for control of seedborne and seed-transmitted pathogens and for health regulation and control by phytosanitary certification and quarantine programs in domestic and international seed trade.
- Seed health testing is critical for insuring the health of basic seed stocks used for seed production and for plant germ plasm utilized in research and product development.
- As a consequence of increased product liability and competitive pressures in seed industry, seed health has become an increasingly important quality trait in marketplace, particularly in vegetable crops.
- Seed sampling is first critical activity in a reliable seed health testing program.
- Seed health tests are always performed on sample of seed taken from seed lot.
- It is, therefore, crucial that the sample used for testing be representative of the seed lot, and this requires standardized sampling procedures.
- Seed testing organizations such as the International Seed Testing Association and the Association of Official Seed Analysts have developed specific rules and procedures for seed sampling for the evaluation of seed quality traits such as germination and physical purity.
- Some countries and some states have included sampling protocols in their seed regulations, and these are generally similar, if not identical, to ISTA or AOSA procedures.
- Seed health testing programs typically utilize the same standardized seed sampling procedures that are used in tests for germination, physical purity, noxious weeds, and other seed and seed lot characteristics.

7.6.3 Symptoms of seed-borne diseases

Viruses

Alfalfa mosaic virus (AMV)

- Plants may develop bright yellow leaf mottle, tip necrosis, reddening, stunting, pod flattening and blackening.
- Yields are reduced through plant death, the production of small seeds, and seeds with brown coat discolouration.
- A high incidence of virus is found in lentil, lupin and chickpeas.

Bean yellow mosaic virus (BYMV)

- Plant leaves may develop leaf mottle or a distinctive yellow mosaic pattern, stunting and reduced leaf size.
- Early infections can seriously reduce plant growth and grain yield.
- It is an economically important virus for lupins.

Cucumber mosaic virus (CMV)

- Plants may develop leaf chlorosis (yellowing), stunting, distortion or bunchy appearance.
- Pods may be flattened and turn purple-brown.
- It causes problems in lentil, lupin and chickpea.

Pea seed-borne mosaic virus (PSbMV)

- Plants may develop downward rolling of leaf margins and slight clearing of the veins in young leaves.
- Production of small seeds with distinctive brown staining and 'tennis ball' marking is common.
- Seed discolouration can significantly reduce the commercial value of grain.
- It is an economically important virus for field peas.

Fungi

Ascochyta blight

- Tan coloured lesions form on leaves, stems and pods.
- Infected leaves may drop prematurely.
- Yields are reduced through plant death and crop damage.
- Infected seed can be shrivelled or badly stained.

Grey mould and chocolate spot (Botrytis cinerea and Botrytis fabae)

- Stem infection can cause the damping-off of young seedlings.
- Later infection causes grey mould or chocolate spot on foliage and flowers.
- Yields are reduced through plant death, crop damage and flower abortion.
- Infected seed can be small and badly discoloured.

Lupin anthracnose (Colletotrichum lupini)

- A severe foliar disease that causes bending and twisting of stems, with a lesion in the crook of the bend.
- Stem infection often results in the death of plants and major yield losses.
- It affects both narrow leaf lupins and albus lupins.

Phomopsis stem blight (Phomopsis leptostromiformis)

- It causes yellow-brown lesions on leaves stems and pods.
- Severe infection can kill plants.
- Infected seed can be covered with a web-like grey mould.
- Toxin produced by infected stubble can kill animals.

Brown leaf spot (Pleiochaeta setosa)

- It cause root rot and leaf spotting.
- In this, affected roots develop large dark brown lesions.
- Irregularly shaped dark brown lesions develop on infected leaves.
- It seriously reduce plant growth and grain yield.

Bacteria

Bacterial blight

- It is an important seed-borne disease of field peas.
- In this, pathogens are carried by seed internally or externally.
- Water-soaked leaf, stem and pod lesions may occur at any growth stage.
- Yields are reduced through plant death, crop damage and the production of small seeds.
- Brown discolouration of the seed coat can sometimes occur.

Epidemiology of seed-borne diseases

Infection of seed

- Seed infection levels are determined primarily by weather conditions between flowering and maturity.
- Warm, humid conditions during this period often results in heavy pod and seed infection.
- Dry weather between flowering and maturity minimises pod infection, and is essential for the production of pathogen-free seed.
- There is often considerable variation between genotypes in their resistance to seed infection.
- Virus infection in seed depends on the amount of spread of virus between plants during the growing season, often by aphid vectors, and the genetic susceptibility of the host plant.

Seed to plant transmission

- Crop losses are likely to be higher for pathogens that invade the roots soon after germination than for pathogens that affect the shoots of young plants.
- The extent of 'seed-plant transmission' of fungal and bacterial pathogens is known to vary considerably depending on infection conditions.
- Poor germination and diseased seedlings can result from the use of infected seed lots.

Survival on seed

- Seed-borne pathogens can often survive for several years in and on seed.

- Some pathogens are carried on the seed coat while others can be harboured deep inside the seed.
- Fungi and bacteria are mostly located on the seed coat, and embryo infection is uncommon.
- Viruses are not carried on the seed coat, and are only found in the seed embryo or tissues of the seed coat.
- Infestation levels of most pathogens decrease rapidly during storage, and long-term storage can eliminate some pathogens from seed.
- Unfortunately, there is likely to be a marked reduction in the viability of seed stored for a long period and this may negate any benefits from lowering disease levels in seed.

Unit 8

Maintain Health and Safety During Seed Production, Processing and Storage

8.1 Perform general safety rules

8.1.1 Idea about injuries due to machinery/vehicle

Farm equipments and vehicles daily check are simple and effective way to spot potentially dangerous issues or defects before they are used. Farmers or seed growers should need to have a system of daily routine checks for machineries and equipments that are used in farms to ensure that are in good working order, safe and fit for purpose before start of working day or work shift.

Farmers and workers use tractors and other equipment every day. These machines are powerful, and if not operated safely, can lead to disabling injuries or death. Modern tractors, telehandlers and other farm equipment have given farmers with an ability to work more quickly and efficiently, but these machinery brings new dangers.

These agricultural machines have potentially dangerous moving parts, which can cause serious or fatal injuries such as:

- balers – pick-ups, twine mechanisms and moving rear doors;
- forage harvesters – chopping cylinders;
- combine harvesters – augers in the grain tank and the header unit;
- potato harvesters – rotating rollers and conveyors;
- slurry tankers – power take-off (PTO) shafts;
- bale and straw choppers – chopping mechanisms;
- tractor – hitch mechanisms, PTOs and PTO shafts;
- power harrows – rotating tines;
- feeder wagons – rotating components in the mixing chamber.

Poorly used or faulty machinery is the main cause of death and injury on farms. Machinery accidents can be prevented by keeping the machine in good repair, fitting and ensuring all safety equipment like guards, safe access platforms and ROPS to be operational with machine all the times.

Farmers and seed growers before using the machines or vehicles should ask themselves:

- Is the machine they intend to use is suitable for the job?
- Are all safety devices like guards are in place and are working correctly?
- Are machine operator properly trained to do this job/use this machine?
- Has the instruction manual for the machine been provided, read and understood?
- Is the right personal protective equipment is available and worn?
- Has a risk assessment been carried out?
- Has the work been properly planned and communicated to those who may be at risk?
- If two or more people are involved, has everyone understood what needs to be done and has a system of communication been agreed on?
- Are all operators/workers trained to do this job?
- Is suitable clothing and footwear available and worn (including any PPE)?
- Have operators been provided with relevant information?

8.1.2 Risk at farms

Risk is a chance that someone could be harmed by hazards like working from ladders or working around electricity. A risk assessment is an important step in protecting people, business and complying with applicable laws.

Farmers and agriculture workers are particularly exposed to ergonomic hazards such as:

- tiring or painful positions
- carrying or moving heavy loads
- standing or walking and repetitive hand or arm movements
- work-related musculoskeletal problems

Farmers live with risk and make decisions every day that affect their farming operations. Many factors that affect the decisions which farmers make could not be predicted accurately such as:

- change in weather conditions
- drop in prices at the time of harvest
- non availability of hired labour during working times
- breakage of machinery and equipment when required
- dying of draught animals
- change in government policies

The above changes are examples of risks which farmers face in managing their farm business. Such risks affect farm profitability.

Sources of Risk

The most common sources of risk in farming can be in areas of:

- Production: Production risk occurs from uncertainty of factors that affects quantity and quality of farm produce.
- Marketing: Marketing risk occurs due to variability of product prices and uncertainty of future market prices.
- Financial: Financial risk occurs when money is borrowed to finance the farm business.
- Institutional: Institutional risk occurs due to unpredictable changes in provision of services.
- Human: Human risk involves risk to farm business that is caused by human illness and personal situation of farmer's family.

Based on farmer's attitude, risk divides farmers among:

Risk-averse farmers

- The risk-averse farmers avoid taking risks.
- They tend to be more cautious individuals with preferences for less risky sources of income.
- They sacrifice certain amount of income to reduce the chance of low income and losses.
- They does not refuse to accept any risk at all.
- The risk averse farmer seek to be compensated for the risk taken by receiving a higher return than would normally be obtained if there were no risk.

Risk-taker farmers

- The risk-taker farmers are people who are open to more risky business options.
- They choose the alternatives which gives some chance of higher outcome, even though they may have to accept lower outcome.
- They are influenced by the return they could receive.

Risk-neutral farmers

- Risk-neutral lies between the risk-averse and risk-taking positions.

Workplace Risk

The main causes of risk/accidents due to equipments while working at site include:

- transport (running over or vehicle over turnings);
- falling from height (from fragile roofs, trees etc);
- struck by moving or falling objects (bales, trees etc);
- trapped by something collapsing or overturning;
- livestock related accidents and fatalities;

- asphyxiation / drowning.

Workplace Risk due to Farm vehicles

Farm vehicles and implements are involved in many farm injuries and fatalities, so number of factors that will contribute to risk, will include:

- unrestrained passengers being carried in the tray or on trailers
- unsupervised children around moving vehicles
- inexperienced drivers
- poorly maintained vehicles.

Vehicle operator should always ensure that:

- passengers use seatbelts
- the driver is trained and licensed
- adopt safe driving methods
- passengers are not carried in the back/tray
- the vehicle is maintained regularly
- children are kept away from driveways and moving vehicles.

Workplace Risk due to Harvester

The hazards due to harvester include:

- entangled with levelling or discharge of augers in grain tank
- falling from machine during pre/post season cleaning
- contacting overhead power lines
- being running over
- contacting with knife, reel, stripper rotor, straw, chopper or spreader
- being trapped under header or injured by header falling from its transport trailer

To avoid risk dont:

- climb or reach in grain tank unless the engine and all augers are stopped, and ignition key has been removed
- climb round areas of the combine unless there is purpose built access available
- carry passengers on header unless passenger seat is provided
- allow people jump on or off from harvester when it is moving
- work under an unsupported header
- carry out maintenance or try to adjust automatic sensors with the engine running
- park, carry out maintenance or extend the unloading auger when under or near over head power lines
- run the combine with guards raised or removed

- allow children on or near the combine
- brake harshly, especially on slopes.

Harvester operator should:

- make sure that anyone working on harvester has read and understood the instruction book
- carry out 'Safe stop' procedure before carrying out maintenance
- have system for thorough preseason maintenance checks to ensure that all brakes, guards and other safety devices are in effective working order
- use any grain sampling mechanism provided
- have safe system of work for combining paddocks where there are overhead power lines.
- use appropriate farm maps as reminders when daily work is being planned
- agree on routes for travel with local company which take into account safe clearance distances
- follow correct procedures for working under header or when transferring header on and off its transport trailer
- provide and use correct tools and knife guard to aid safe removal, handling and storage of the cutting knives
- work with cab door shut
- be particularly careful when reversing
- wear slip resistant footwear and avoid wearing loose clothing
- take care on steep ground and avoid sudden changes of direction.
- park across slopes if possible and ensure handbrake is fully applied before leaving the vehicle
- make sure all guards are in position and correctly fitted before starting work
- regularly clean straw and chaff deposits from the engine compartment and around belts and pulleys
- make sure there is suitable fire-fighting equipment available on the combine

General safety

- Make sure all guards and shields are in place.
- Ensure all workers are trained in proper operating technique and safety procedures when using machinery.
- Stop all moving parts before servicing the machine.
- Operators should take regular breaks.
- Protective footwear and close-fitting clothes are essential when working in and around machinery.

- ❖ Use suitable access equipment and take precautions for work at height during cleaning and maintenance operations.
- ❖ Be patient when working in difficult crop conditions such as laid crops or unfamiliar fibrous crops like linseed.
- ❖ Don't operate the machine beyond its capacity.
- ❖ Use your instruments and watch and listen for potential blockages – avoiding blockages is easier than clearing them.
- ❖ Use drive reversing mechanisms, when fitted, to wind out blockages.
- ❖ Remember the baler operator – try to avoid leaving heaps in the swath.
- ❖ Remember the hazards posed by straw choppers and spreaders – allow adequate rundown time before approaching the rear of the combine.
- ❖ Combines fitted with yield measuring meters using an ionising radiation source need additional safety procedures which must be laid down, clearly understood and implemented.
- ❖ Exposure to high levels of grain dust causes ill health, including occupational asthma, farmer's lung, grain fever, chronic bronchitis, allergic eye and nasal infections.
- ❖ Make sure the cab filter is of the type recommended by the manufacturer and is fitted and maintained in accordance with the manufacturer's instructions.
- ❖ When unloading on the move consider the other driver and avoid sudden manoeuvres.

Managing Risk

Reducing risk are often used in combination with one another as no single strategy can cover all of risk likely to be encountered. Farmers need to consider the risks simultaneously and to develop an integrated approach for better management. They need to recognize the advantages and disadvantages of each risk management option both individually and in combination. Each farmers should select an appropriate strategy based on their goals, attitudes towards risk and their personal and financial situations.

Farmers should choose and combine strategies based on their goals, attitudes towards risk and their personal and financial situations. Every farmer family required to find their own ways of coping with risk. Farmers' responses to risk are as diverse as the risks that affect their farms.

Farmers can reduce risk by learning about and applying new technologies and practices designed to address specific risks common to their area of production such as new varieties of seed are being developed and livestock are being bred with certain characteristics like:

- ❖ drought-resistant seed for wheat;
- ❖ bird-resistant seed for sorghum;
- ❖ disease- and pest-resistant seed species;
- ❖ disease-resistant livestock species;
- ❖ livestock bred to provide better productivity;

- irrigation for high-value crops;
- crops and livestock bred specifically to improve marketability.

Factors involved in managing risk

Avoiding Production Risk

- Selecting low-risk activities is a way to reduce production risk where farmers choose reliability over potential profitability.
- In this, farmer may forego an enterprise that has a high potential for income but also carries high risk for loss, and choose instead an enterprise which is less profitable but also less risky.

Using Flexible Farming

- Farming system flexibility is an important strategy for managing risk.
- A flexible farming system makes it possible for farmer to make quick or short-term changes in production and sales.
- Farmers who sell cash crops also reduce risk by using available funds to enable them to change to other enterprise if price of main cash crop falls.
- By keeping farm systems flexible, farmers are able to make decisions in response to changing circumstances.
- If farmers work on general production plans, they should keep their options as open as possible in order to respond to opportunities and risks as they occur.

Production diversification

- Diversification spreads risk and is successful risk management strategy as not all farm enterprises and operations are likely to be affected in the same way by changing situations. It involves:
- Managing multiple farm enterprises together at any one time (or in the same season);
- Engaging in the same farm enterprise in different physical locations;
- Engaging in the same farm enterprise over successive periods of time (or seasons);
- Generating income from off-farm activities.

Engaging in same enterprise in different physical locations

- Such type of risk managing strategy works on understanding that the same crops grown in different areas will not meet the same fate.
- Perhaps a crop grown in the area surrounding the house is more likely to be infested by rodents than the same crop grown elsewhere.

8.2 Gain knowledge of various health hazards relevant to work place and basic first aid training

8.2.1 Understanding health hazards by use of herbicides/pesticides/fumigants

Pesticides

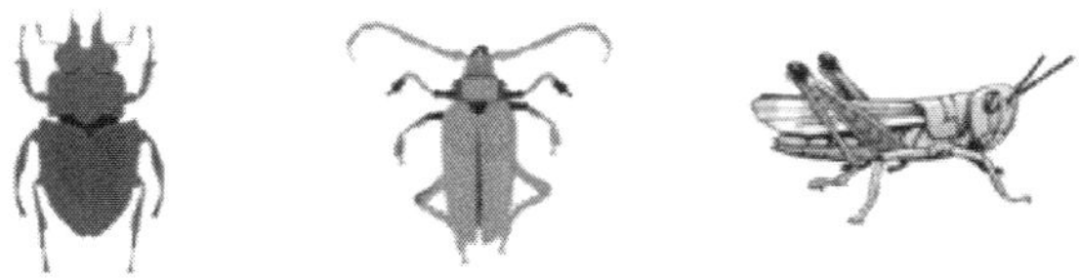

Pesticide is a group which contains:

- ❖ insecticides,
- ❖ fungicides,
- ❖ herbicides,
- ❖ rodenticides,
- ❖ garden chemicals,
- ❖ wood preservatives
- ❖ household disinfectants

Pesticides have been an essential part of agriculture as they protect crops and livestock from pest infestations and gives reduction for many decades. It is poisons and can be hazardous.

Farmers and seed growers are exposed to pesticides in many different situations such as in:

- Mixing
- Loading
- Spraying
- Storing
- Spillage
- Disposal

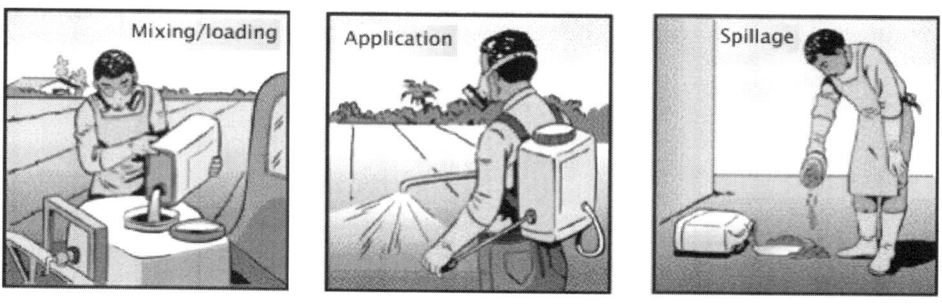

Pesticides enter in body through:

- ❖ Skin
- ❖ Breathing
- ❖ Mouth

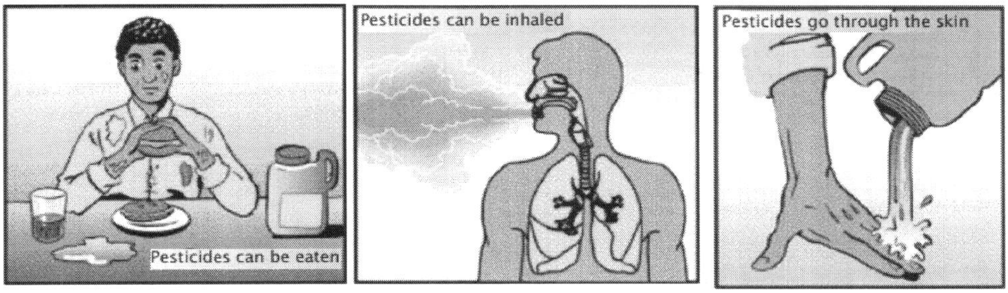

Fumigants

- ❖ Fumigation is pest control method where pesticide gas or vapor is released in air or injected in the soil to kill or eliminate pests.
- ❖ It fills the airspace in a structure with toxic gas.
- ❖ It uses tarp or tent to trap the gas in the area which requires treatment.
- ❖ It is commonly used in agricultural practice to treat fields to reduce weeds, nematodes and other soil insects and small rodents such as moles.
- ❖ In this, the fumigants are injected in soil and move through the soil air which dissolve in the water in the soil where they kill pests.
- ❖ Fumigant pesticides are used on grains and crops post-harvest to reduce insects, ticks and mites, nematodes, slugs and snails and fungal diseases.

Since pesticides, herbicides and fumigation requires specialized agrichemicals, so farmers or farm workers should know different hazards and should monitor to see different risk level. Farmers or agricultural workers are exposed to chemicals like pesticides, veterinary drugs, solvents, and oils that causes different long-term hazards to humans.

The effects of chemical exposure to farmers or farm workers while working on fields depend on the type of chemical and degree of exposure. If chemicals are swallowed, absorbed through the skin or inhaled as mist, vapour or dust, then there appears a different immediate effect such as:

- ❖ poisoning
- ❖ headache
- ❖ nausea
- ❖ vomiting
- ❖ diarrhoea
- ❖ pinpoint pupils

- dizziness
- fine muscle twitching
- increased bronchial and lacrimal secretions
- skin rashes and irritation
- chemical burns.

8.2.2 Provide first aid trainings

While working at site, seed growers and labours involves in certain type injuries. It is the duty of gardener/supervisor to keep/arrange first aid tools at site, so that in case of any emergency, an immediate first aid could be given to the patients using first aid techniques.

First aid kits and attendants

- Make every worker aware of where the first aid kit is located and how to call the first aid attendant if one is required.
- Post signs indicating how to access first aid.
- If a first aid attendant is required, make sure the attendant holds a first aid certificate of the level necessary for that workplace.
- If you require a first aid attendant, train backup attendants. Ensure that enough workers are trained for this responsibility to cover vacations and other absences.

Common site injuries could be bleeding, burns, eye injury, chemicals inside the eyes wounds and breathing etc.

Control of bleeding

- Direct pressure – use your hand(s).
- Elevate (raise) the limb
- Apply a pad and firm bandage.
- If necessary use clean rags or clothing.

Remember

- Always check circulation below the bandage!
- If there is tingling, numbness or blueness loosen the bandage.

Burns

- Cool the burnt area with cool water for 10-15 minutes.
- If necessary, cover the burn with a clean dressing or plastic wrap before removing person to medical aid.

Remember

- Do not burst blisters.
- Do not remove clothing that is stuck.
- Do not apply creams.

Eye injuries

Foreign bodies in the eye(s)

- ❖ Wash the eye(s) with eyewash or clean water.
- ❖ If the foreign body is stuck to the eye surface, do not attempt to remove it.
- ❖ Place a covering over the eye(s) and send for, or take, the person to medical aid.

Chemicals in eye(s)

- ❖ Wash the eye(s) with clean cool water for at least 15 minutes.
- ❖ Wash from near the nose outwards and always wash under the upper eyelid.
- ❖ Send for, or take, person to medical aid.

Minor wounds

Abrasions, small cuts

1. Clean the wound with mild antiseptic or soap and water.
2. Cover lightly with clean dressing.

Breathing

If a person is breathing but unconscious, turn them onto their side to prevent airway obstruction (tongue or vomit).

Handling/storing of first aid equipment

Immediately and properly examining and treating of injuries save life and is essential to reduce pain and help injured people make a quick recovery. All this is possible at site through first aid kit which should comprise of following:

- ❖ Band-Aids
- ❖ Absorbent compress dressings
- ❖ Sterile gauze pads
- ❖ Adhesive cloth tape
- ❖ Gloves
- ❖ Roller bandages
- ❖ Antiseptic wipes
- ❖ Antibiotic ointment
- ❖ Scissors and tweezers
- ❖ Sterile saline
- ❖ Hydrocortisone ointment

It is important to check the first aid kit on regular basis. Medications expire, so you may need to replace old ones. You need to replenish supplies that you have used in the past, so make sure that more common treatments such as medications, bandages etc are always available. If possible use antiseptic solution to clean the scissors and injections.

Reporting first aid procedure at workplace

Gardener/seed grower must provide information about first aid to their workers so that they should know what to do and whom to contact in case of any emergency. Information should be easy to understand, accessible and should take into account the language and literacy levels of the workers. Information may be given using verbal methods or visual methods. The information and instruction on first aid should include:

- location of first aid equipment and facilities
- names and location of persons trained to administer first aid
- procedures to be followed when first aid is required.

The information and instruction should be provided as part of workers' induction training and when there are any changes, for example in the location of first aid facilities or in the names, locations or contact details of first aiders.

It is the duty of the gardener/seed grower to check that medical person should keep all the details regarding the type of treatment, what he/she is giving along with details of first aid equipment which are dispensed. Medical officers should record maximum information about the patient and type of injury, what the patient is suffering from, with old patient history. This information is important when handing-over the casualty to a health care professional which should cover:

- Name and contact number of casualty;
- Nature of the injury;
- Time and date of incident;
- Details of treatment given

8.3 Understand basic safety checks and other common reported hazards before all operations at field and stores.

8.3.1 Workplace safety

Creating safe workplace is an ongoing commitment. Employers/seed grower can make the company a safe place by doing the following or ensuring that it is done (for example, by a supervisor):

- Provide regular safety training for supervisors and workers.
- Lead by example and demonstrate safe work habits.
- Regularly check that workers are following company safety procedures.
- Regularly check that workers are correctly using the protective equipment and devices provided.
- Encourage workers to report illnesses and injuries immediately.
- Encourage workers to report potential hazards.
- Keep workplace health and safety programs current.
- Stay aware of workplace hazards and how to handle them.

❖ Respond promptly to all health and safety concerns.

8.3.2 Health and safety checks in workplace

There are certain health and safety procedures which should be checked and followed by gardener/seed grower while working in area such as:

1. Dry and clean slip proof floors/pathways:

❖ Slips and falls are one of the common ways of enduring injuries, so to prevent injuries, the floors/pathways should be cleaned on regular basis.

❖ While planting seed, mud or water mixture creates slip, so soon after the planting of seeds, the working area should be clean.

❖ Slip can also occur due to wet contaminants such as water, oil or grease, or dry such as dust, metal, plastic bags or off-cuts.

❖ Preventing floor contaminants is one of the best things you can do to prevent slips.

❖ More serious slips or trips together with the resulting falls may result in:

- sprains or strains
- broken bones when trying to break the fall
- a back injury due to the sudden and forceful impact during a fall
- burns if it occurs near hot surfaces or if the person is handling hot fluids
- cuts if it occurs near sharp objects.

2. Unorganized and messy environment

❖ The messy environment is the main cause of all problems.

❖ In this, there are more chances of threats going unnoticeable and the situation may simply turn the place into accident-prone area.

❖ An occupational hazard is a result of ignorance on part of workers or carelessness on part of workers.

- ❖ If workplace is messy and unorganized, it may be a door to many uninvited hazardous accidents, so a gardener should properly arrange and clean the working area from tools, chemicals, fertilizers etc.
- ❖ Proper piling of pots, fertilizer bags, and chemical boxes should be done in the store area, otherwise locating the items from improper piling will lead to stack falling upon workers and can make him/her injured.
- ❖ An unorganized store may also hamper the productivity rate of the worker as for locating particular item from the store without proper stocking will lead to time wastage of worker.

3. Use of disinfectants in workplace

- ❖ Disinfectants are very important to be used in garden, parks, public places as there are more risk of germs spreading around due to multiple uses of each and every item.
- ❖ In workplaces, there is frequent use of items from one worker to another, whether it is flu season or not, as germs are inevitable in nature.
- ❖ The disinfectants comparatively lessen the chance of germ produced illness or diseases.
- ❖ In order to cut the cost some workers opt for cheap disinfectant, which on longer run can affect the health and in turn, will lead to health issues and absentees.

4. Germ controls to fight flu/diseases:

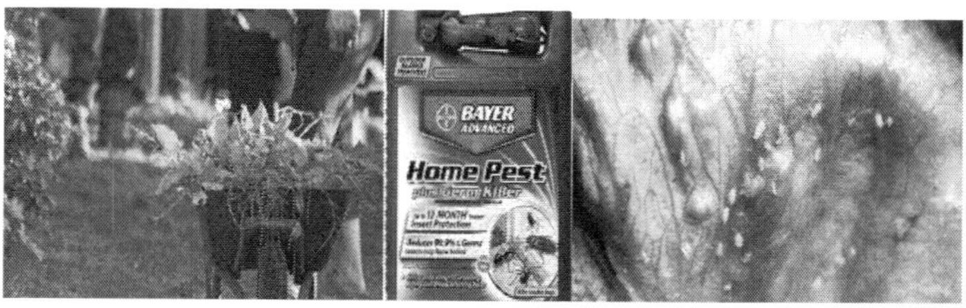

- ❖ Germs can easily disperse across the entire workplace if there is any compromise in the cleanliness.
- ❖ The sitting place or lunchroom can be called as germ house for workers, so it is important to clean and sanitize such areas frequently.
- ❖ Apart from flu, there is high chance of catching fatal disease or infection by shared germs in the workplace.

❖ There are many infectious germs that may spread due to unhygienic environment and lead to various disease such as fungal infection which can lead to irritation and skin infections.

5. Green and clean equipments/products

❖ Eco-friendly products are best in nature and are not harmful to human health.

❖ Such products are friendly to nature.

❖ Not all cleaning products can be compared to one another as all such products have different standards.

❖ Some products are cheaper and harmful whereas some are costlier and trustworthy.

❖ It is very important to choose the correct chemical.

❖ While choosing chemicals, it is equally important to store these chemicals in correct places.

❖ There are more chances of such chemical to cause skin related allergies, which on longer run can affect the worker's attention span and confidence.

6. Air filtering and humidity

❖ Airborne diseases do not stay back in race of harming and spreading more and more diseases.

❖ One may not be able see them, but they can be equal threat in the way of the organization at a whole.

❖ A better ventilation system works as backbone against airborne diseases.

❖ Germs that can pass through respiratory infections can be tackled by ventilating system.

❖ It will not only enhance the productivity of workers but also maintain their good health.

- ❖ Thus, regular cleaning and timely changing of the required filters is very much advisable. Germs restrict the potential microbial growth or odor related problems in work environment.
- ❖ Humidity affect the germ level in workplace, so maintaining proper humidity level in store room or sitting area is advisable.

7. Proper lightening

- ❖ Dirty light fixtures are indeed one reason to inefficient light supply.
- ❖ It can be a great hassle in way of workers.
- ❖ It is proven that adequate light is required in order to do daily task.
- ❖ No proper lightening can be an unsafe combination with workplace.
- ❖ Garden lawns, pathways, sore rooms and equipment areas should be well lighted to prevent accidents and contribute to healthy and safe environment.

8. Perfect waste disposal plan

- ❖ Piled up trash will result in dirt all around the workplaces.
- ❖ Piled trash is a place where pesticides can reside and multiply.
- ❖ Pesticides can be a great threat and hazard to working environment.
- ❖ Placing dust bins or waste bins throughout the workplaces is a way to gather waste in a single place.
- ❖ Proper collection of waste from dustbins should be followed so that the waste will not spread anywhere in working area.
- ❖ Gardener/worker should implement reducing and recycling methods in workplace.

9. Image is everything

- ❖ Workplace should be clean as apart from workers, there are customers, suppliers or sellers who used to roam in workplace.
- ❖ If the environment of workplace is messed up, unorganized and dirty, then there will be less possibility of good customers to visit and buy the product.
- ❖ Clean workplace will give better impression to the customers and will lead to good product sale.

8.4 Understand, identify and study the use of equipments, processing machines and materials safely and correctly

8.4.1 Understand and check machine before use

Before working with any machinery, farmer/seed grower should carry out the basic check to make sure that the machine/vehicle what he/she will be using is in the good working order and is safe for use. For this, the machine/vehicle should be checked with basic checking like:

- ❖ Mechanical defects – pay particular attention to items like brakes, wheels and tyres.
- ❖ Guards and other protective devices should be correctly fitted and maintained in good condition.
- ❖ Check that Guards and other protective devices are in correct place, securely attached and working properly.
- ❖ The damaged or defective guards/devices should be repaired or replaced before the machine is used.
- ❖ Check that stopping devices are functioning correctly, such as emergency stops.
- ❖ See that all controls should be clearly marked to show what they do.
- ❖ If workers are to be carried on the machine ensure that they can do so safely, eg check for safe means of access and that working platforms have guard rails etc.
- ❖ Hitching and attachment points – check that mounted, semi-mounted or trailed machinery are safely attached to the towing vehicle such as a tractor.
- ❖ Examine the conditions of drawbar/pick-up hitch, hitch rings, pins, clips etc.
- ❖ Carry out any pre-use checks as specified in the operator's manual.

- ❖ For self-propelled machines, make sure that mirrors are clean and properly adjusted.
- ❖ Check any other reversing aids are working.

Checking Harvesting Equipment

Farmer/seed grower should check for following before using grain harvesting equipment:

- ❖ Before trying to operate a combine, operator/farmer should study the operator's manual which carries information about all general safety rules with specific safety recommendations for particular machine.
- ❖ Farmer/operator if checking the combine inside the building should make sure that all doors are opened to provide good ventilation as exhaust fumes from gasoline or diesel engine are poisonous which can lead to series health hazard.
- ❖ Always clean the combine before starting as trash around the exhaust system can cause fires.
- ❖ Clean the oil, grease or mud on ladders or on platform of the combine as they can cause serious falls.
- ❖ Check if combine is equipped with cab, then clean the glass to provide maximum visibility.
- ❖ Check the tire pressure of combine every day as under-inflation can lead to buckling of sidewall, which causes dangerous tire failure.
- ❖ Check the brakes once in a week as with hydraulic brakes, it is required that the master cylinder should have full fluid and should be free from air inside the lines.
- ❖ Check the threshing cylinder-rocking bar to see it is clear of the cylinder.
- ❖ Make sure that all shields and covers are in place and fastened securely.
- ❖ Remove or stow all service equipment.
- ❖ Always use the handrails and ladders provided on the combine for safe mounting and dismounting.
- ❖ Be sure that all PTO covers, safety stands and shields are on the machine before taking it to the field.

Checking Bailing Hay Equipment:

Farmer/seed grower should check for following before using the bailing hay equipment:

- ❖ Check that the baler should be properly connected to the tractor.
- ❖ Bailer should have required safety guards fitted.
- ❖ See that nobody should be allowed to ride on the baler.
- ❖ See that nobody is getting too close to the baler.
- ❖ Check that fire extinguisher should be fitted to the machine.
- ❖ Disengage PTO and apply fly wheel brake before doing any baler adjustments.
- ❖ Stop engine and apply fly wheel brake before repairs or stringing up the baler.

8.5 Understand and handle the emergency situation in field and store operations

8.5.1 Introduction

Emergency procedure is a plan of actions which needs to be done in certain order or manner, in response to specific class of foreseeable emergency, which poses immediate risk to health, life, property, or environment.

There are many emergencies which require urgent intervention to prevent a worsening of situation, while in some situations, mitigation may not be possible and agencies will offer only palliative care for aftermath. In such case, emergency plan should allow for certain possibilities.

An emergency procedure should provide with appropriate and proportionate response to certain situation and should identify responsibilities, actions and resources required to deal with emergency situation.

Once everything has been drafted, a procedure should be followed by consultative period with those who involved/affected by emergency, and a programme should be set out for testing, training and periodic review.

8.5.2 Workplace emergency

Emergencies at workplace include:

- Floods
- Fires
- Toxic gas releases
- Chemical spills
- Radiological accidents
- Explosions
- Civil disturbances
- Workplace violence

8.5.3 Emergency response plans

Nurseries/fields should be prepared to respond to emergencies such as fires, chemical spills, or natural disasters. If an emergency occurs, there will be a need to make quick decisions that will minimize injuries and damage. Such decisions are easier if seed grower have already developed an emergency plan.

Follow these guidelines to develop and implement a plan:

- List all possible events (for example, serious injuries, fires, explosions, or natural disasters).
- Identify the major consequences associated with each event (for example, casualties, equipment damage, or facility damage).

- ❖ Determine the necessary measures to deal with those consequences (for example, first aid, notification of medical authorities, rescue, firefighting, or equipment evacuation).
- ❖ Determine what resources will be required (for example, medical supplies or rescue equipment).
- ❖ Store emergency equipment where it will be accessible in case of emergency.
- ❖ Ensure that workers are trained in emergency procedures and shown where equipment is stored.
- ❖ Hold periodic drills at least annually to ensure that employees will be ready to act if an emergency occurs.
- ❖ Communicate the plan to everyone involved.

8.5.4 Role of gardener during emergency/evacuation

- ❖ Assessing the situation to determine whether an emergency exists requiring activation of the emergency procedures;
- ❖ Supervising all efforts in the area, including evacuating personnel;
- ❖ Coordinating outside emergency services, such as medical aid and local fire departments, and ensuring their available and notified when necessary;
- ❖ Directing the shutdown of workplace operations when required.

8.5.5 Fire and emergency procedures

It is the duty of gardener/seed grower to know and understand the following procedures in case of fire:

1. Raising the alarm and evacuating people from the work area
2. Calling emergency services
3. Calling the manager/incharge
4. In case of small fire, use fire extinguisher to extinguish the fire
5. If the fire is near a fuel tank, do not extinguish the fire, but retreat to safe distance
6. Be prepared to direct the fire service to the fire area

At workplace, emergency procedures depend on the worksite which includes:

- ❖ Emergency shutdown
- ❖ Evacuation
- ❖ First aid
- ❖ Fire fighting

1. Emergency Shutdown of Equipment

- ❖ If there is a fire, emergency or accident, in such case you need to stop using the equipments.

2. Emergency evacuation procedures

- ❖ On hearing an evacuation alarm, or on instruction of emergency control personnel, immediately cease all activity and safeguard personal valuables.
- ❖ Help any person in immediate danger, but only if required to do so.
- ❖ Safeguard any activity/process that results hazardous or suffer damage if left unattended as a consequence of evacuation.
- ❖ Act as per directions given by emergency control personnel and evacuate the work area immediately.
- ❖ Assist with general evacuation if directed to do so by emergency control personnel.
- ❖ Assist with evacuation of disabled occupants.
- ❖ In fire, do not try to work in fields.
- ❖ Move calmly to required evacuation area and do not leave the evacuation area till everything has been cleared.
- ❖ Follow instructions of relevant emergency services personnel.

3. First Aid

- ❖ First Aid is the quick care given to an injured or ill person.
- ❖ Every site has a first aid officer.
- ❖ If somebody needs first aid, he/she must tell their supervisor or manager.

4. Fire Fighting Equipment

- ❖ Fire fighting equipment on site could be anything from small fire extinguishers through to large water cannons.
- ❖ Different fire fighting equipment should be used for different types of fire.
- ❖ Always check the equipment for information on what type of fire it can be used on.

Steps for using a fire extinguisher:

- ❖ Evacuate the area.
- ❖ Isolate the area.
- ❖ Call emergency services or other designated on site procedure.

- If it is safe to do so, use an extinguisher to attempt to control the fire.

8.5.6 Spill procedures

It is the duty of gardener/seed grower to know and understand the following procedure in case of occurrence of spill:

- If the spill is from the hose or tap, shut the isolation valve
- Warn people in area of the spill and evacuate, if necessary
- Remove sources of ignition if flammable substances are there
- Weigh up the spill by only response if think it is safe to do
- Refer to the safety data sheet or call on an approved handler or other specialists for advice
- If necessary, call emergency services and take advise from local authority
- Put on safety equipment (boots, gloves, eye protection, etc.)
- Contain the spill if it is safe to do so using drip tray or oversize container or spill kit to soak the substance
- Dispose of waste safely

8.5.7. Using emergency action plan

An emergency action plan covers required actions which workers and seed grower must take to ensure safety from fire and other emergencies.

Emergency action plan must include:

- A required method for reporting fires and other emergencies;
- An evacuation policy and procedure;
- Emergency escape procedures and route assignments like floor plans, workplace maps, and safe or refuge areas;
- Details of individuals both within and outside the workplace to contact for additional information/explanation of duties and responsibilities under the emergency plan;
- Procedures for workers who remain to perform or shut down critical workplace operations, operate fire extinguishers or performing other essential services which cannot be shut down for every emergency alarm before evacuating;
- Rescue and medical duties for any workers designated to perform them.

Glossary

'A' line: cytoplasmic male sterile seed parent used in making commercial hybrids.

Admixture: something added to seed other than the kind and/or variety specified

Apomixis: reproduction from an unfertilized eggs or from somatic cells associated with the egg.

Andromonoecious: refers to a plant species in which male and bisexual flowers are produced on the same plant e.g., muskmelon and ridgegourd.

'B' line: (i) fertile counterpart of the 'A' line. The 'B' line does not have fertility-restoring genes and is used as the male parent to maintain the 'A' line, i.e., 'A' line x 'B' line reproduces the 'A' line. The 'B' line is fertile and can be reproduced by self-fertilization;

Berry: a simple fleshy fruit.

Bulb: a short, underground stem with fleshy leaves

Bulk: assembly of similar materials from a collection into one or more bulks.

Border rows: the recommended number of rows of the male parental line to be grown on all the sides of the seed fields growing two different parents.

Chromosomes: small structures in the nucleus of a cell that carry the genes. They appear as thread or rod-shaped structures during metaphase. Each species has a characteristic number of chromosomes.

Clone: a group of individuals of common ancestry which have been propagated vegetatively (asexually), usually by cutting or natural multiplication of bulbs or tubers.

Commercial hybrid: refers to the first generation F1 of the hybrid planted for any purpose.

Companion cropping: some crops are the best companions and when together they complete their life cycles in equal period.

Composite: a composite is developed by intermating (three to four generations) selected open-pollinated parent varieties or lines known for their diversity. Parents are selected on the basis of their per se performance or their general combining ability. Such constructed composite populations are made for use in recurrent selecting programmes.

Container: a box, bottle, casket, tin, barrel, case, receptacle, sack, bag, wrapper or other things in which any article or thing is placed or packed.

Corn: a bulky, short, vertical, underground stem which stores food.

Cross-fertilization: union of an egg with a sperm from a plant of a different clone.

Cross-pollination: transfer of pollen from an anther to the stigma in a flower on a different plant.

Crown: persistent base of a tufted perennial herbaceous plant as found in grasses and asparagus.

Curd: mass of flower primordium meristems as found in cauliflower (heading broccoli).

Cyme: centrifugal inflorescence on which the secondary or lateral branches continue to grow and extend beyond main axis.

Cytoplasm: main contents of a cell in which the nucleus and other bodies are located.

Cytoplasmic-male sterility: a type of male sterility conditioned by the cytoplasm rather than by nuclear genes and transmitted only through female parent.

Diploid: organism or cell with two sets of chromosomes.

Detassel: to remove the tassel or pollen producing organ at the top of maize plant before pollen is released.

Dioecious: having staminate and pistillate flowers on different plants of the same species for example spinach, hops, hemp, date palm, papaya, pointed gourd, little gourd, asparagus etc.

Emasculation: removal of stamens before they burst and shed their pollen.

Ergot: fungus disease of cereals and grasses which forms a sclerotium in place of a healthy grain, often poisonous to man and animals.

F1: generation that arises from a given crossing; filial generation.

F2: generation produced by selfing the F1; the second filial generation.

Field inspection: official inspection of seed fields conducted by the official of a Certification Agency or his authorized agent.

Fodder crop: crop grown for feeding to animals.

Forage: in range management, unharvested plant material of any kind available for animal consumption. It may be used for grazing or be cut for feeding. When cut, it becomes feed (hay or silage).

Full-sib: term used in population improvement. A full-sib family comprises progeny from a cross between two selected plants within the population.

Gall: swelling or excrescence of tissue of plants resulting from the attacks of certain parasites or seed structures in which the contents have been replaced by nematodes.

Gene pool: useful genes or gene complexes in a divergent population.

Generation: one complete life cycle. The generation begins with the formation of the zygote and end when the resulting plant dies.

Genetic sterility: type of male sterility conditioned by nuclear genes. In contrast to cytoplasmic sterility, it may be transmitted by either the male or female parent.

Germination: in a laboratory test, the emergence and development from the seed embryo of those essential structure which for the kind of seed being tested, indicate the ability to develop into a normal plant under favourable conditions in soil.

Growing season: period (s) of the year during which crops grow and mature.

Grow-out test: test performed to determine the genuineness of seed as to species or variety.

Gynomonoecious: refers to a plant species in which female and bisexual flowers are produced on the same plant e.g., cucumber.

Hardseeds: seeds that have a seed coat impervious to water or oxygen required for germination for example; seeds of Leguminosae, Malvaceae etc.

Hectare: standard area measure in the metric system, 10,000 square meters in area.

Heterozygous: having unlike alleles at corresponding loci of homologous chromosomes. An organism may be heterozygous for one or several genes.

Homozygous: having genes at corresponding loci on homologous chromosomes that are identical.

Hybrid: plant resulting from a cross between parents that are genetically unlike.

Hybridization: crossing of one plant with another. Crossing between plants of the same species is called intraspecific hybridization and crossing between different species is called interspecific hybridization.

Infection: at the epidemiological level of the pathosystem, infection refers to the contact made between host and parasite; hence, auto-infection, allo-infection. At the histological level, infection refers to the process of penetration of a host by a pathogen.

Infest: to over run the surface of a plant, or to be dispersed through soil or other substrate.

Inert matter: seed like structures from crop and weed plants that are one half the original size or less; badly injured and undeveloped seed like structures of weeds, glumes, stems and other parts; sand dirt and other related substances.

Inseparable other crop: crop whose seeds are difficult to be separated once mixed with the main crop seed.

Inter-cropping: when two crops of different height, canopy, adaptation, root system and growth habit are made to grow simultaneously in such a way that they accommodate each other with least competition.

Isolation: the act of keeping the seed crops away from the sources of physical and genetical contamination.

Isolation distance: distance to be maintained between the seed crop and the contaminant.

Kind: one or more related species or sub-species of crop plants each individually or collectively known by one common name such as cabbage, maize, paddy and wheat.

Male sterile: describes the complete or partial failure of a male plant to produce mature reproductive pollen cells.

Mixed cropping: two or more than two crops are grown in mixed stand either as 'birra' mixture or in separate rows in different proportion. When any of the crop inmixed stand gets vitiated due to one reason or other, the other crop may act as an insurance against complete loss or other.

Moisture: amount of water present in seeds.

Monoecious: having staminate and pistillate flowers on the same plant. For example corn; cucurbits etc.

Multiline variety: consist of two or more near isogenic lines of normally self-fertilizing plants which are similar in most characteristics but differ in a limited number of describable physiological, morphological or other essential or distinctive characteristics. A multiline is derived by growing the component lines separately compositing the lines to constitute the breeder class of seed.

Nematode: thread like round worms which live in soil and water but also live in plants.

Objectionable weed: weed whose seeds are difficult to be separated once mixed with crop seed or which are poisonous or injurious or are having smothering effect on the main crop or are difficult to eradicate once established or are having high multiplication ratio thus making their spread quick.

Offtype/rogue: plant which does not conform to the varietal characteristics described by the breeder. To designate a plant as an offtype/ rogue it is not necessary to identify it as to variety.

Other crop seeds: seeds of plants grown as crops.

Outcross: mating of a hybrid with a third parent; also an offtype plant resulting from pollen of a different sort contaminating a seed field.

Panicle: in a cereal crop, that portion of the plant that bears seeds.

Phytotoxic: Poisonous to plants

Planting ratio: the recommended ratio in which the male and female parental lines are to be panted to make a crossing block in hybrid seed production

Planting stake: stem cutting used for planting.

Pod: dry and many seeded dehiscent fruit e.g., as found in Leguminosae or Cruciferae.

Pollen: male gamete or a plant that is produced in the anthers

Pollen parent: parent that furnishes the pollen which fertilizes the ovule of the other parent (seed parent) in the production of seed.

Pollen shedders: plants of 'B' line which is male fertile when present in 'A' line which is male sterile.

Pollinator: line or population used as a male parent (pollen donor).

Pollination: transfer of pollen from the anther to the stigma. Pollination must occur before fertilization can take place.

Polyploid: plant having other than the diploid (2n) number of chromosomes.

Pure seed: seed of all botanical varieties of each species under analysis.

Producer: a person who grows or distributes certified seed in accordance with the procedures and standards of the Certification Agency.

Previous crop: crop of the same kind grown in the crop season immediately preceding the one in which the seed crops is grown.

Previous season: crop season immediately preceding the one in which the seed crop is grown.

Ratoon crop: crop obtained from re-growth from stubble (living stumps) following a harvest, not necessarily of grain. For example; sugarcane, sorghum, rice, millets, oats, pigeon pea etc.

Ratooning: cutting a plant to obtain re-growth.

Raw seed: harvested seed that has not been cleaned and graded.

'R' line (Restorer line):an inbred line that when crossed to the A-line (male sterile strain) cause the resulting hybrid to be male fertile and produces pollen.

Recessive gene:(i) a gene not expressed in the heterozygous state when a dominant gene is present at the same locus on the other homologous chromosome; (ii) a gene masked by the effect of another specifically, a gene the effects of which are masked by a dominant allele.

Receptive silks: silks which are fresh, green/yellow/pink (typical to code designation/ variety) and not dried. When more than one cob/ear on a plant have receptive silk it is counted as one only.

Rhizomes: a horizontal underground stem frequently a storage organ.

Root slips: the plant plants which arise from portion of root when it is bedded for propagation.

Sclerotia: compact tissue like, sometimes rounded mass of hyphae (threads of fungal mycelium), with or without the addition of host tissue, normally not having spores in or on it e.g. ergot.

Seed: (i) a mature ovule, consisting of an embryonic plant together with a store of food all surrounded by a protective coat. It usually develops after the fertilization on an egg cell by male generative cell from a pollen grain. Seeds of some species develop without the intervention of the male cell, formed entirely of 'mother' tissue, such seeds are called apogamic seeds.

(ii) "seed" means any of the following classes of seeds used for sowing or planting:

Seed parent: strain from which seed is harvested in the hybrid seed field.

Seed borne disease: disease which is carried either within the seed i.e. internally seed borne or on the seed i.e. externally seed borne or both.

Selfing: plant breeding term usually meaning natural or artificial self-pollination.

Self-fertilization: union of an egg with a sperm from the same flower or from another flower on the same plant or within a clone.

Self-pollination: transfer of pollen from an anther to the stigma of the same flower or another flower on the same plant or within a clone.

Sibbing: **sib-mating:** Cross between plants from the same population. Generally pollen is collected from several plants from the same population, bulked and crossed onto sister plants of the same population.

Smut balls: mass of smut spores, a disease usually of cereals.

Species: a unit in botanical classification, a sub-division of a genus. A grouped of closely related individuals descendant from the same stock.

Spikelet: separate flower cluster in grasses, consisting usually of two sterile glumes at the base of one or more florets.

Sterile: not fertile, unable to reproduce sexually.

Strain: a group of similar individuals within a variety. (i) advanced generation random-mating population derived from a few selected inbred lines; (ii) a variety produced by crossing inter-se a number of inbred lines (usually five to eight) selected for their good general combining ability. The variety is subsequently maintained by open-pollination.

Test cross: cross made with a homozygous recessive parent to determine whether an individual is homozygous or heterozygous.

Tetraploid: polyploidy plant having four sets of identic similar chromosomes (4n).

Tiller: branch arising from the base of a monocot plant, especially in the grass family.

Triploid: organism or cell with three sets of chromosomes (3n).

Tuber: an enlarged portion of a rhizome (underground stem) that stores food.

Umbel: inflorescence in which the stalks of the flowers all spring from the top of the main stock, like the ribs of an umbrella, e.g., in unbelliferae.

Variety: a sub-division of a kind identifiable by growth, yield, plant, fruit, seed or other characteristics. It also denotes on assemblage of cultivated individuals which are distinguished by a character significant for the purposes of agriculture, or horticulture and which when reproduced or reconstituted retain their distinguishing features.

Vapour-proof container: container which does not permit entrance or exist of moisture.

Viable: alive, able to live, grow and develop.

Viviparous: germinating or sprouting from seed or bud while still attached to the parent plant e.g., chow-chow.

Volunteer plant: plant of the crop species same as that of the seed crop which comes up on account of self-seeding from the previous season's crop.

Weed: plant out of place. Any plant growing where not wanted.

Weed seed: seed, bulblet, tuber of plants recognized as weed by laws, official regulations or by general usage.

Wild types: naturally occurring non-domesticated crop relatives.

Index

O

P

Q

R

S

T

V

W